PROBLEMS AND SOLUTIONS FOR INTEGER AND COMBINATORIAL OPTIMIZATION

T0315423

Problems and Solutions for Integer and Combinatorial Optimization

Building Skills in Discrete Optimization

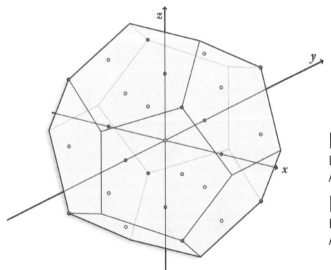

Mustafa Ç. Pınar
Bilkent University
Ankara, Turkey

Deniz Akkaya
Bilkent University
Ankara, Turkey

Society for Industrial and Applied Mathematics
Philadelphia

Mathematical Optimization Society
Philadelphia

Publications Director	Kivmars H. Bowling
Executive Editor	Elizabeth Greenspan
Acquisitions Editor	Paula Callaghan
Developmental Editor	Rose Kolassiba
Managing Editor	Kelly Thomas
Production Editor	Ann Manning Allen
Copy Editor	Ann Manning Allen
Production Coordinator	Cally A. Shrader
Compositor	Cheryl Hufnagle
Graphic Designer	Doug Smock

Library of Congress Control Number: 2023944156

 is a registered trademark.

 Mathematical Optimization Society is a registered trademark.

Dedicated to the memory of Mustafa Pınar's mother,
Suzan (1934-2023),
and
his uncle, Özalp Gökbilgin (1935-2021).

Dedicated to Deniz Akkaya's mother, Özden, his father,
Murat, and his brother, Alp Akkaya.

Contents

II   Solutions   63

1   Mathematical Modeling Problems with Integer Variables   65

List of Figures

List of Tables

List of Symbols

$\mathbb{B}$ the set $\{0, 1\}$

$\mathbb{F}_2$ the finite field of two elements, namely $\{0, 1\}$

$\mathbb{N}$ the set of natural numbers

$\mathbb{Z}$ the set of integers

$\mathbb{Q}$ the set of rational numbers

$\mathbb{R}$ the set of real numbers

A_+ the non-negative subset of A, i.e., $A \cap [0, \infty)$

A_{++} the positive subset of A, i.e., $A \cap (0, \infty)$

$|A|$ cardinality of A

A^n the set of all n-dimensional vectors with entries from A, i.e., $\underbrace{A \times A \times \cdots \times A}_{n \text{ copies}}$

$A^{m \times n}$ the set of all $m \times n$-dimensional matrices with entries from A, i.e., $\underbrace{A^m \times A^m \times \cdots \times A^m}_{n \text{ copies}}$

LHS left-hand side

RHS right-hand side

WLOG without loss of generality

$>>$ much greater than

$\mathbf{x}$ a vector

x_i ith entry of vector $\mathbf{x}$

y_{ij} entry at ith row and jth column of matrix Y

Preface

The authors of this book wanted for a long time to have at hand a large number of problems with solutions while teaching. The first author has taught IE303 Modeling and Methods in Optimization for Bilkent University students for longer than 20 years. A graduate student (a PhD candidate) assistant joined next and started compiling solved problems to be used in exams, quizzes, and homework assignments. The first author added some problems of his own. The result is this book. There are not many books out there dedicated to problems in integer optimization and related topics. The book focuses on the topics covered in IE303 Modeling and Methods in Optimization, a third year required course for Industrial Engineering students at Bilkent University. However, it should be useful for any undergraduate student in industrial engineering or in related disciplines. These are the motivations behind the preparation of this book.

The readers are advised to avoid checking out the solutions immediately and work on their own to figure out a solution. After all, mathematics is not a spectator sport. You have to get your hands dirty. And nothing equals the pleasure of conquest.

Topics covered in the aforementioned course are modeling capabilities of integer variables, the Branch-and-Bound method, cutting planes, network optimization models, shortest path problems, optimum tree problems, maximal cardinality matching problems, matching-covering duality, symmetric and asymmetric TSP, 2-matching and 1-tree relaxations, VRP formulations, and dynamic programming. These are topics covered in most undergraduate level textbooks such as Winston [25] or Taha [22].

The problems in this book cover a plethora of topics and applications. They vary in scope and in difficulty although the majority have short answers (they may have taken longer to state!). Some of them introduce concepts and techniques that may not have been discussed at an undergraduate level on optimization. Instructors may also develop extensions of some exercises as they see fit.

There are occasionally a few problems which may require the use of a modeling system such as GAMS, Gurobi, XPRESS-MP, or Excel solver or even the use of a programming language.

Structure of the book

The book has three units: each unit focuses on one particular topic. Chapter 1 covers mostly modeling aspects of integer optimization in various situations. Chapter 2 is about some problems which are "easy" at least under some assumptions, but not so easy in general. Chapter 3 focuses on techniques used to deal with integer optimization problems or how to solve them, along with other applications.

How to use this book

The selection of problems to assign to students is left at the discretion of the instructor making use of the book. We envisage two ways the instructor can use this book: the first is to keep the book out of reach of students and assign problems to be graded from the book. This method is not our preferred method since students will quickly discover the source. The second method that we would recommend is to share the book with students (they may be asked to buy a copy) and assign problems to be discussed in problem sessions and recitations. This second method is preferable since in our experience most of the time students feel the need to discuss the solutions at length even if they have access to written solutions which may be too succinct for their tastes.

About the companion website

The soon to be developed website (https://bookstore.siam.org/mo33/bonus) for this book will contain

- errata and corrections;

- new problems that were written after the publication of the current version;

- miscellaneous material (e.g., suggested readings, etc.).

Acknowledgments

We are deeply indebted to colleagues and friends for their contributions which found their way into this project, and two anonymous reviewers for their very detailed suggestions.

Mustafa Ç. Pınar and Deniz Akkaya
Bilkent University

Part I

Problems

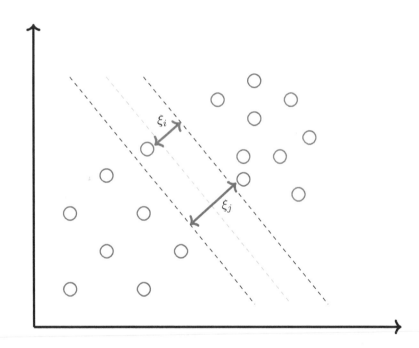

Chapter 1

Mathematical Modeling Problems with Integer Variables

In this chapter we gathered problems that help illustrate the capabilities offered by integer variables for dealing with various requirements in numerous fields of mathematics and several engineeing applications, e.g., logical circuits (see Figure 1). The reader or instructor can refer to [24] for a detailed study of mathematical models in integer optimization. Another source with various example integer programming models is the book [9].

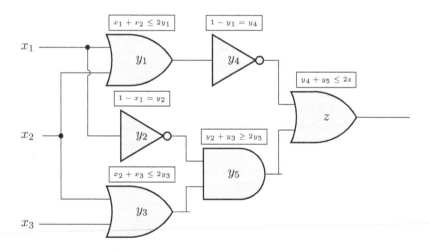

Figure 1. *A Logic Circuit Explained with Binary Variables.*

1.1 ▪ Greatest Common Divisor

The greatest common divisor of two integers $a_0, a_1 \in \mathbb{Z}$, where not both a_0 and a_1 are zero, is the largest integer that divides both a_0 and a_1. It is denoted by $\gcd(a_0, a_1)$. The following is well known. The greatest common divisor $\gcd(a, b)$ is the smallest integer in the set

$$\left\{ ax + by : \begin{array}{l} ax + by \geq 1, \\ x, y \in \mathbb{Z}. \end{array} \right\}$$

Give an integer linear programming formulation for the computation of the greatest common divisor of two integers.

1.2 ▪ Bin Packing

Suppose that a set of items $\mathcal{N} = \{1, \ldots, n\}$ is given. Item $i \in \mathcal{N}$ has size a_i, which is a positive integer. We would like to put the items into bins. Each bin has size b, which is also a positive integer. Suppose n bins are available and $a_i \leq b$ for all $i \in \mathcal{N}$. Each item should be put into exactly one bin, and the total size of items put in a bin should not exceed the size of the bin. The aim of the problem is to minimize the maximum empty space left in the used bins.

Let x_{ij} be 1 if item $i \in \mathcal{N}$ is put into bin $j \in \mathcal{N}$ and 0 otherwise. Let y_j be 1 if bin $j \in \mathcal{N}$ is used and 0 otherwise. Let z be the maximum empty space left. Using these variables formulates an integer programming model for this problem.

1.3 ▪ Three Cups Puzzle

Suppose you have three water glasses, named A, B, and C. Given the initial position of the problem, the objective of the Three Cups Puzzle is to turn all cups right-side up in no more than six moves, turning over exactly two cups at each move. Formulate an integer program to solve the Three Cups Puzzle with the initial position as below making as few moves as possible.

1.4 ▪ Table Tennis Tournament

Suppose you are arranging a four-day table tennis tournament consisting of two teams with four members each. Each day there are four time blocks, 10 am, 12 am, 2 pm, and 4 pm. Arrange a schedule such that

- each player plays against all members of the opponent team;
- at each time block there will be a single game;
- a player does not play twice in the same time block or day.

Formulate an integer programming model to solve this problem.

1.5 ▪ Floor and Ceiling Operators

In integer optimization, there are some commonly used operators which are floor ($\lfloor \cdot \rfloor$) and ceiling ($\lceil \cdot \rceil$). They are defined as follows:

$$\lfloor x \rfloor := \text{largest integer smaller than or equal to } x,$$
$$\lceil x \rceil := \text{smallest integer greater than or equal to } x.$$

Find a linear mixed integer program equivalent to the following non-linear program:

$$\begin{aligned}
\text{maximize} \quad & x + y \\
\text{subject to} \quad & \lfloor x \rfloor + \lceil y \rceil = 5, \\
& x, y \geq -5.
\end{aligned} \tag{P}$$

Hint: You may use $a + \epsilon \leq b$ instead of $a < b$ for sufficiently small $\epsilon > 0$.

1.6 ▪ A Chemical Reaction

Suppose you are given the following chemical equation:

$$x_1 KNO_3 + x_2 C \rightarrow x_3 K_2CO_3 + x_4 CO + x_5 N_2,$$

where the LHS of the arrow consists of reactants and the RHS consists of products.

(a) Formulate an integer programming program to find the appropriate positive integral coefficients $\{x_1, \ldots, x_5\}$ with minimum total sum such that the number of atoms and the mass are conserved, i.e., the number of atoms in reactants should be equal to products.

(b) Is the optimal solution unique? Either prove it or find two alternative optima.

Hint: Null space of homogeneous equality constraints may help.

1.7 ▪ Largest k Elements of a Vector

In the following problems you have $\{v_1, \ldots, v_n\}$ positive integers, and k is also a positive integer with $k < n$.

(a) Suppose you have a vector $v = \begin{bmatrix} v_1 & v_2 & \cdots & v_n \end{bmatrix}^T$. Formulate an integer linear program to find the largest k elements of the vector using an inequality instead of an equality.

(b) Suppose Ömer is preparing for the university entrance exam. For n days, his teacher assigned him v_i questions for each day i. Ömer may choose the number of questions he wants to solve each day, say y_i, but the teacher penalizes for unsolved questions. On day $n+1$ he has to solve the following number of questions:

$$k \max\{v_1 - y_1, v_2 - y_2, \ldots, v_n - y_n, 0\}.$$

Formulate an integer linear program for Ömer to solve the smallest number of questions possible in $n + 1$ days.

(c) Show that linear relaxations of the formulations in **(a)** and **(b)** have primal-dual relation.

Hint: Use the fact that problems $\max \mathbf{c}^T \mathbf{x}$ st. $A\mathbf{x} \le \mathbf{b}, \mathbf{x} \ge \mathbf{0}$, and $\min \mathbf{b}^T \mathbf{y}$ st. $A^T\mathbf{y} \ge \mathbf{c}, \mathbf{y} \ge \mathbf{0}$, have primal-dual relation.

1.8 ▪ Coloring Nodes of a Graph and Classroom Assignment

The YouLearn University Department of Philosophy is trying to schedule the summer session courses. Having to figure out how many classrooms they will need, they made a list of sections i in $\mathcal{L}$. For each course section i, a list $N(i)$ contains a list of sections taught by the same instructor, while another list $T(i)$ contains the list of sections that have been scheduled at the same time. Obviously, a section i and the sections $j \in N(i) \cup T(i)$ cannot be scheduled in the same classroom. By likening the problem to a node coloring problem, create an associated graph by clearly defining the nodes and edges of the graph. Then, give an integer programming formulation that minimizes the number of distinct classrooms needed to schedule all courses of the department.

1.9 ▪ A Chessboard Problem

Suppose you have an 8×8 regular chessboard, and you have an unlimited number of 2×1 domino tiles.

(a) Formulate an integer program to find the maximum number of domino tiles that can be placed on the chessboard. Solve it by inspection.

(b) Suppose your chessboard has two diagonally opposite corners removed, leaving 62 squares. Solve the same problem, without formulating, by inspection.

1.10 ▪ The 8-Queens

The *Queens Patrol Problem* is the problem of placing the maximum possible number of queens on an 8×8 chessboard (see Figure 2) such that none of them threatens another (no two are in the same row, column, or diagonal). Formulate an integer program to solve this problem.

Figure 2. *A Feasible Solution for 8-Queens.*

1.11 ▪ A Basket of Apples

Suppose you have m apples with weights w_i and two baskets with capacity C. Find an arrangement of the apples in a way that the difference between the total weights in the baskets is minimized. What would you do if we had more than two baskets, each basket had a different capacity C_i, and you wished to minimize the maximum difference between the weights of the baskets?

1.12 ▪ A Grocery Chain

A grocery chain is deciding to open up new stores. Unfortunately vegetable products are highly vulnerable to rot. Therefore it is highly important to decide the locations of the stores to be able to serve each customer while avoiding leftovers at the end of the day. The fixed cost of opening a store at location i is B_i. The opened shops will be serving customers immediately. The cost of serving a product from the ith shop to the jth customer is denoted as c_{ij}. There are a few important issues to consider:

- No inventory is held at all times as food tends to go bad.

- Each store i has the capacity to serve s_i units.

- Unsatisfied demand results in customer dissatisfaction, and the cost associated for that unmet demand is p_j.

(a) Model the problem above as a linear integer program, assuming there are n stores and m customers which are $m, n \gg 8$ and all coefficients are integers.

(b) There has to be a store opened at location 1 or location 4. Also, the number of total stores opened should not exceed 6.

(c) If stores are opened at locations 1, 3, and 5, then both stores at locations 7 and 8 should be opened.

1.13 ▪ An Infeasible Integer Program

Suppose you are asked to solve the following linear program:

$$\text{minimize} \quad \mathbf{c}^T \mathbf{x}$$
$$\mathbf{a}_i^T \mathbf{x} \le b_i, \quad \forall i \in \{1, \dots, m\},$$
$$x_j \le 1, \qquad \forall j \in \{1, \dots, n\},$$
$$\mathbf{x} \in \mathbb{R}_+^n,$$

and it turned out to be an infeasible problem. Suppose you can delete some of the inequalities of the type $\mathbf{a}_i^T \mathbf{x} \le b_i$ in order to obtain a feasible problem, but your objective function increases by $d \gg c_j$ each time you delete one. Formulate a mixed integer program to find a minimum cost feasible problem. If you use the big-M method, clearly write what big-M is.

1.14 ▪ Packaging Chicken Nuggets

ReMO is a fast-food company selling chicken nuggets. They sell nuggets in packs of six, nine, and twenty. One customer figured out that while he can buy exactly 21 nuggets $(6 + 6 + 9)$, he cannot buy exactly 16 nuggets using the provided package sizes. There is an optimum integer number k^* such that the company can deliver all orders of size bigger than k^*, i.e., sizes $k^* + 1, k^* + 2, k^* + 3, \dots$.
Give an integer programming model to find k^* with available packaging of 6, 9, and 20 nuggets.

Hint: If the package sizes were 3 and 5, then the answer would be 7 since any order with size larger than 7 can be served by using these packages. E.g., an order for 8 nuggets is delivered using one packet of 3 and one of 5, an order for 9 nuggets is delivered with 3 packages of size 3, and so on.

You may end up with a model having infinitely many constraints. In that case you need to find a finite set of constraints describing an equivalent feasible region.

1.15 ▪ Handling Unions of Polyhedra

Let two polyhedra

$$P_1 = \left\{ (x, y) \in \mathbb{R}^2 : \begin{array}{l} 2 \le x \le 4, \\ 0 \le y \le 1, \end{array} \right\}$$

and

$$P_2 = \left\{ (x, y) \in \mathbb{R}^2 : \begin{array}{l} 3 \le x \le 5, \\ 3 \le y \le 4, \end{array} \right\}$$

be given. Suppose we have the following optimization problem:

$$\text{minimize} \quad 3x + 2y$$
$$\text{subject to} \quad (x, y) \in P_1 \cup P_2.$$

Reformulate the problem above as a mixed integer linear program.

1.16 ▪ Simplifying an Integer Program

Let $a, b \geq 0$. Suppose we have the following integer linear program:

$$\text{maximize} \quad ax_1 + ax_2 + bx_3$$
$$\text{subject to} \quad x_1 + x_2 + x_3 \leq \tfrac{3}{2},$$
$$-x_1 - x_2 - x_3 \geq -\tfrac{17}{10},$$
$$x_1 + x_3 \leq \tfrac{13}{10},$$
$$-x_1 \leq 0,$$
$$-x_2 \leq 0,$$
$$x_3 \leq \tfrac{11}{5},$$
$$x_1 + x_2 + 2x_3 \leq \tfrac{12}{5},$$
$$\mathbf{x} \in \mathbb{Z}^3.$$

Simplify the problem above as much as possible and obtain an equivalent optimization problem with two variables. For what values of a and b is the problem unbounded?

1.17 ▪ Equivalence of Integer Sets

Let G, H be subsets of $\mathbb{R}^n$. If $\mathbb{Z}^n \cap G = \mathbb{Z}^n \cap H$, we say G and H are *equivalent integer regions*, $G \sim H$.

Suppose we have the set $H = \{(x, y) \in \mathbb{R}^2 : (x^2 + y^2 - 1)^3 \leq x^2 y^3\}$.

(a) Find the smallest set $G \in \mathbb{R}^2$ such that $H \sim G$.

(b) Find the smallest polyhedron P such that $H \cap P$ is the smallest convex set such that $H \sim H \cap P$.

(c) Solve

$$\text{maximize} \quad 3x - 4y$$
$$\text{subject to} \quad (x, y) \in H,$$
$$x, y \in \mathbb{Z}.$$

Hint: It is difficult to plot H, but it is a heart-shaped region around the origin. You may start by showing $x^2 + y^2 \leq 2$ for any $(x, y) \in H$.

1.18 ▪ Generalized Assignment Problem

Objects $i \in \{1, \ldots, 10\}$ of volume c_i cubic meters are being stored in an automated warehouse. Storage locations $j \in \{1, 2, 3\}$ are located d_j meters from the input/output station, and all have a capacity b cubic meters.

(a) Formulate an optimization model to store *all the items* at minimum total travel distance assuming that as many objects can be placed in any location as volume permits and objects cannot be divided. Define your decision variables and constraints clearly.

(b) After the initial formulation, management expressed the following requirement: if items 1 and 2 are stored in warehouse 1, then items 3 and 5 should also be stored in warehouse 1. Formulate two constraints for this requirement.

Hint: The following rules are helpful: $\delta = 1 \Rightarrow \sum_{j=1}^{n} a_j x_j \geq b$ requires $\sum_{j \in N} a_j x_j + m\delta \geq m + b$, and $\sum_{j=1}^{n} a_j x_j \geq b \Rightarrow \delta = 1$ requires $\sum_{j=1}^{n} a_j x_j - (M + \epsilon)\delta \leq b - \epsilon$. M is an upper bound on $\sum_{j=1}^{n} a_j x_j - b$, whereas m is a lower bound on the same quantity. (Take $\epsilon = 1$.) See [20, 24] for these rules.

1.19 ▪ Basketball Team Line-up

A basketball coach is trying to choose the starting line-up for the basketball team. The team consists of seven players who have been rated (on a scale of 1 = poor to 3 = excellent) according to their ball-handling, shooting, rebounding, and defensive abilities. The positions that each player is allowed to play and the players' abilities are listed below.

Player	Position	Ball-handling	Shooting	Rebounding	Defense
1	G	3	3	1	3
2	C	2	1	3	2
3	G-F	2	3	2	2
4	F-C	1	3	3	1
5	G-F	3	3	3	3
6	F-C	3	1	2	3
7	G-F	3	2	2	1

The five-player starting line-up must satisfy the following restrictions:

- At least 3 members must be able to play guard, at least 2 members must be able to play forward, and at least 1 member must be able to play center.

- The *average* ball-handling, shooting, and rebounding level of the starting line-up must be at least 2.

- If player 3 starts, then player 6 cannot start.

- If player 1 starts, then players 4 and 5 must both start.

- Either player 2 or player 3 must start.

Given these constraints, the coach wants to maximize the total defensive ability of the starting team. Formulate an integer program that will help him to choose his starting team.

1.20 ▪ Non-linearities

Let (P) be a non-linear integer program defined as

$$\begin{aligned}
\text{minimize} \quad & 2x + 3y \\
\text{subject to} \quad & |x| + |y| \leq 3, \\
& x, y \in \mathbb{Z}.
\end{aligned} \tag{P}$$

(a) Rewrite (P) as an integer linear program without defining extra variables.

(b) Rewrite (P) as an integer linear program using extra variables.

1.21 ▪ Product of Two Binary Variables

Consider the following requirement, which arises in an integer programming formulation problem: the binary variables δ_1, δ_2, and δ_3 should satisfy $\delta_1 \delta_2 = \delta_3$. Note that the value of $\delta_1 \delta_2$ is also binary. Write a system of linear inequalities that is equivalent to the given requirement.

1.22 ▪ Sorting

Let $\mathbf{x} \in \mathbb{R}^n$ and $M \geq 0$. We define *the sort function* as follows:

$$\text{sort}(\mathbf{x})[i] = i\text{th smallest element in } \mathbf{x}.$$

For example let $\mathbf{x} = \begin{bmatrix} 1 \\ 3 \\ 2 \end{bmatrix}$. Then $\text{sort}(\mathbf{x}) = \begin{bmatrix} 1 \\ 2 \\ 3 \end{bmatrix}$.

Suppose we have the following optimization problem:

$$
\begin{aligned}
\text{maximize} \quad & \mathbf{c}^T\mathbf{x} + \mathbf{f}^T\mathbf{y} \\
\text{subject to} \quad & A\mathbf{x} \leq \mathbf{b}, \\
& |x_i| \leq M, \qquad \forall i \in \{1, \ldots, n\}, \\
& \mathbf{y} = \text{sort}(\mathbf{x}), \\
& \mathbf{x}, \mathbf{y} \in \mathbb{R}^N.
\end{aligned} \qquad (P)
$$

Formulate a mixed integer linear program equivalent to (P).

1.23 ▪ Magic Squares

A *magic square* is an $n \times n$ square grid (where n is the number of cells on each side) filled with distinct positive integers in the range $\{1, \ldots, n^2\}$ such that each cell contains a different integer and the sum of the integers in each row, column, and diagonal is equal. The sum is called the *magic constant* of the magic square. Formulate an integer programming problem to find a magic square of order n.

1.24 ▪ Lights-out Puzzle or the Game of Fiver

The lights-out puzzle is played on 5×5 board where each cell is on or off. Initially, all the cells are on. One can click on a cell with the result that it turns off and all four of its neighbors north, south, east, and west switch status, i.e., if they are on they are turned off. However, if a cell is clicked a second time (or a neighbor is clicked), the cell changes status (from on to off or vice versa). The game consists of turning a fully on board fully off by making the smallest number of cell clicks possible. Give an integer linear programming model to solve the puzzle.

1.25 ▪ Sparse Coding

Considering a matrix $Y = [\mathbf{y}_1, \ldots, \mathbf{y}_i, \ldots, \mathbf{y}_\ell] \in \mathbb{R}^{n \times \ell}$ of signals of dimension n, a sparse representation of Y consists in finding a matrix $X = [\mathbf{x}_1, \ldots, \mathbf{x}_\ell] \in \mathbb{R}^{p \times \ell}$ of decomposition coefficients, which is sparse over a learned dictionary $D = [\mathbf{d}_1, \ldots, \mathbf{d}_p] \in \mathbb{R}^{n \times p}$. The columns of the latter, i.e., $\mathbf{d}_j$ for $j \in \{1, \ldots, p\}$, are called atoms. Sparse representations have been considered with success in signal and image processing.

Since the dictionary D is fixed, the sparse coefficients $\mathbf{x}$ for each signal y can be obtained by solving the following problem:

$$
\begin{aligned}
\text{minimize} \quad & \|\mathbf{y} - D\mathbf{x}\|_2^2 \\
\text{subject to} \quad & \|\mathbf{x}\|_0 \leq T, \\
& \mathbf{x} \in \mathbb{R}^p,
\end{aligned} \qquad (P)
$$

where the notation $\|\mathbf{x}\|_0$ refers to the number of non-zero elements of the vector $\mathbf{x}$. This formulation is practical for solving the problem with prior knowledge of the sparsity level T of the signals. Give a mixed integer quadratic programming formulation of the problem.

1.26 ▪ Partition to Eliminate Triangles

Model the following feasibility problem as an integer programming feasibility problem: Given a graph $G = (V, E)$, does there exist a partition of the edges E into two sets E_1 and E_2 such that neither E_1 nor E_2 contains a triangle?

1.27 ▪ A Production Planning Problem

An engineering plant can produce five types of products $\{p_1, p_2, \ldots, p_5\}$ by using two production processes: grinding and drilling. Each product requires the following number of hours of each process and contributes the following amount (in hundreds of dollars) to the net total profit:

	p_1	p_2	p_3	p_4	p_5
Grinding	12	20	0	25	15
Drilling	10	8	16	0	0
Profit	55	60	35	40	20

Each unit of each product takes 20 man-hours for final assembly. The factory has 3 grinding machines and two drilling machines. The factory works a six-day week with two shifts of 8 hours/day. Eight workers are employed in assembly, each working one shift per day.

(a) Assuming integer units of production, formulate an integer optimization model to plan the weekly production to maximize total net profit.

(b) Management voiced the following extra requirement: If we manufacture p_1 or p_2 or both, then at least one of p_3, p_4, p_5 must also be manufactured. Incorporate this requirement into your model.

1.28 ▪ Telecommunications Network Design

In a mobile phone network, each basic geographical area, referred to as a *cell*, is covered by a transmitter-receiver device known as a *relay*. All mobile phone calls must initially pass through these relays. These relays are interconnected to a central node called a "hub" using either cables or electromagnetic waves. Among these hubs, one is designated as the primary control center, known as the "MTSO" or Mobile Telephone

Switching Office. These hubs and the MTSO are connected through a ring network using fiber optic links. This network is designed to be robust; it is capable of restoring itself in case of a breakdown without the need for replication.

For the problem at hand, we assume that there are no dynamic connections between the relays and the MTSO. Instead, the connections are pre-determined during the network design phase. Therefore, it is crucial to carefully select the nodes on the ring that connect to a relay. The number of connections between a cell denoted as c and the ring is referred to as the cell's *diversity*, represented as d_c. A higher diversity value is preferred as it enhances network resilience. For instance, if a cell has a diversity of 2, it means it is connected to at least two nodes on the ring.

In this system, all traffic is digitized and measured in values equivalent to bidirectional circuits at 64kbps (kilobits per second), representing the capacity for simultaneous calls during peak usage. The ring's edges have known capacity denoted as U. The traffic from a cell c, denoted as T_c, is evenly distributed among its connections to the ring, resulting in each connection handling $\left(\frac{T_c}{d_c}\right)$ traffic. This traffic is then transmitted through the ring to the MTSO, which further routes it to another cell or to a hub serving as the interface to the fixed-line telephone network. Additionally, a relay may have a direct connection to the MTSO, which can also function as an ordinary hub.

Let's define C as the set of cells in the network. We are dealing with a network comprising $|C|$ cells and a ring with M nodes, each having a capacity of U circuits. The MTSO is located at node i^*. We are given information about traffic demands, the required number of connections, and the cost per connection, denoted as co_{cj} (in thousands of dollars per cell), for connecting cell c to node j.

The objective is to determine the optimal connections between the cells and the ring while minimizing the connection costs, while adhering to the capacity constraints of the ring and satisfying the specified limitations on the number of connections.

Formulate this problem as an integer programming problem by defining decision variables, constraints, and an objective function that optimize the network configuration and minimize costs.

1.29 ▪ Combinatorial Auctions: Winner Determination

In a combinatorial auction, there is a set of items I which are simultaneously for sale, and bidders place bids on bundles. For example, if $I = \{A, B, C, D\}$, a bid might be $(\{A, C, D\}, 5)$, indicating that this bidder is willing to pay 5 for receiving all of the items A, C, and D. In a sealed-bid combinatorial auction, all of these bids are submitted simultaneously. Once the authority/firm organizing the auction (the auctioneer) has received all the bids, it is faced with the *winner determination* problem of deciding which of these bids win. One can award each item at most once, so one cannot accept two bids if they have one or more items in common. Under this constraint, generally, the goal is to accept bids in a way that maximizes the sum of the values of the accepted bids. For example, suppose that there are 4 items, A, B, C, D, and that the auctioneer receives the following bids: $(\{A,B\},4),(\{B,C\},5),(\{A,C\},4),(\{A,B,D\},7),(\{D\},1)$. It is easy to see that of the first 4 bids, one can accept at most 1, because every pair of the first 4 bids overlaps. Hence, the optimal solution is to just accept (A, B, D, 7) and let C go unallocated, for a total value of 7. The second-best feasible solution is to accept (B, C, 5) and (D, 1), for a total value of 6. Formulate the winner determination problem as an integer optimization problem.

1.30 ▪ Combinatorial Reverse Auctions

An important variant of combinatorial auctions is *combinatorial reverse auctions*. In a combinatorial reverse auction, there is a single buyer who needs to procure a set of items I, and there are multiple sellers who are bidding to sell subsets of these items to the buyer. For example, if $I = \{A,B,C,D\}$, a bid might be $(\{A,C,D\};5)$, indicating that the corresponding seller is willing to sell all of A, C, D for a price of 5. It should be noted that unlike in combinatorial (forward) auctions, now, lower bids are better. The buyer must accept a subset of the bids that includes at least one copy of each item (duplicates can be thrown away). Formulate the problem of the buyer who wishes to minimize the total expenditure as an integer optimization problem.

1.31 ▪ Rank Aggregation: Kemeny Rule

Suppose that we have some set of elements and several rankings of these elements. For example, we may have $I = \{A,B,C,D\}$ and the following three rankings: $A \succ_1 B \succ_1 D \succ_1 C$, $C \succ_2 A \succ_2 B \succ_2 D$, and $D \succ_3 B \succ_3 C \succ_3 A$. Here, $a \succ_j b$ indicates that ranking j ranks a higher than b. We wish to obtain a single aggregate ranking $\succ$ of I from these rankings. There are many motivations for this problem. A group of people may have to make a joint choice from a set of alternatives I; the ranking $\succ_j$ corresponds to the jth person's preferences over the alternatives, and the aggregate ranking would correspond to the group's aggregate preferences. (This type of setting is typically referred to as a *social choice* or *voting* setting.) Alternatively, perhaps one entered the same query in several different search engines; each engine j returns its own ranking $\succ_j$ of the relevant pages, and one wishes to come up with an aggregate ranking of these pages. There are many different ways to determine an aggregate ranking, and which one is optimal has been a topic of intense debate in the social choice community at least since the 18th century. Here, we adopt a nice rule for determining an aggregate ranking, the Kemeny rule. The Kemeny rule works as follows.

Given a potential aggregate ranking $\succ$, one of the input rankings $\succ_j$, and $a, b \in I$, let $\delta(\succ; \succ_j; a; b) = 1$ if $a \succ b$ and $b \succ_j a$, and let it be 0 otherwise. Then the *Kemeny score* of ranking $\succ$ is equal to

$$\sum_{j=1}^{n} \sum_{\substack{a,b \in I \\ a \neq b}} \delta(\succ; \succ_j; a; b).$$

That is, the Kemeny score of a potential aggregate ranking is the total number of disagreements, where a disagreement occurs whenever there is some input ranking and some pair of elements such that the aggregate ranking and the input ranking disagree on that pair of elements. The Kemeny rule chooses an aggregate ranking with the lowest Kemeny score (if there are multiple such rankings, the Kemeny rule does not specify a tiebreaker). For the example above, the unique Kemeny ranking is $A \succ B \succ D \succ C$. The score of this ranking is 7 (for every pair of alternatives, there is one disagreement, except for the pair A, C, for which there are two disagreements).

Formulate the problem of finding the optimal Kemeny ranking as an integer optimization problem.

1.32 ▪ Coding Theory and Integer Optimization

How can we transmit information reliably over an *error-prone* transmission system? This problem is at the heart of the mathematical discipline known as *information theory*, and in particular of *coding theory*, which deals with the design of error-correcting codes for the purposes of reliable transmission. In coding theory, the principle of error-correction coding is to preventively include *redundancy* in the transferred messages, thus communicating more than just the actual information, with the goal of enabling the receiver to recover that information, even in the presence of noise on the transmission channel. The general system has the following characteristics:

- The function by which these "bloated" messages are computed from the original ones is called the *encoder*;

- the channel introduces noise to the transmitted signal, i.e., at the receiver there is *uncertainty* about what was actually sent;

- the *decoder* tries to recover the original information from the received signal.

In this project we are interested only in *block codes*, which means that the information enters the encoder in the form of chunks of uniform size (the *information words*), each of which is encoded into a unique coded message of again uniform (but larger) size (the *codewords*). We shall deal with the binary case, i.e., both information words and codewords are binary valued. That is, we have sequences of zeros and ones.

Definition. An (n, k) code is a subset $\mathcal{C} \in \mathbb{F}_2^n$ of cardinality 2^k. A bijective map from $\mathbb{F}_2^k$ is called an encoding function for $\mathcal{C}$.

By a slight abuse of terminology we use $\mathcal{C}$ for both the code and the encoding function.

Consider information words with size $k = 4$. There are 16 possible information words. For instance, Hamming(7,4) is a *linear* error-correcting code that encodes four bits of data into seven bits by adding three parity bits (see https://en.wikipedia.org/wiki/Hamming(7,4) for more information on Hamming codes and background information; it is required reading.)

The numbers k and n are referred to as *information length* and *block length*, respectively. The quotient $r = \frac{k}{n} < 1$ represents the amount of information per coded bit and is called the *rate* of $\mathcal{C}$.

The concept of redundancy is entailed by the fact that $\mathcal{C}$ is usually a strict subset of the space $\mathbb{F}_2^n$, i.e., most of the vectors in $\mathbb{F}_2^n$ are *not* codewords (in fact, we shall be dealing with linear codes in this project; see below). In the present project, the structure of the set of codewords is more important than the actual coding function.

Definition. A binary (n, k) code $\mathcal{C}$ is called *linear* if $\mathcal{C}$ is a linear subspace of $\mathbb{F}_2^n$ (Check the definition of a a linear subspace.) Consequently, a linear code has a linear encoding function.

Linearity of codes allows one to define codes compactly using matrix algebra. All operations on matrices below are performed in binary arithmetic, i.e., "modulo 2."

Definition. A matrix $H \in \mathbb{F}_2^{m \times n}$ is called a *parity-check* matrix if its rows generate $\mathcal{C}^\perp$, the orthogonal complement space $\mathcal{C}^\perp$ (incidentally, $\mathcal{C}^\perp$ is called the code *dual* to $\mathcal{C}$) to $\mathcal{C}$:

$$\mathcal{C}^\perp = \{\boldsymbol{\xi} \in \mathbb{F}_2^n : \boldsymbol{\xi}^T \mathbf{x} = 0 \text{ for all } \mathbf{x} \in \mathcal{C}\}.$$

Based on this definition, the linear code $\mathcal{C}$ is completely characterized by the null-space (or, kernel) of the matrix H:

$$\mathcal{C} = \{\mathbf{x} \in \mathbb{F}_2^n : H\mathbf{x} = \mathbf{0}\}. \tag{1}$$

(Recall that all operations are modulo 2.) The reason why H is called a "parity-check" matrix is explained in https://en.wikipedia.org/wiki/Hamming(7,4) Section 4.

For the code Hamming(7,4) the parity-check matrix H is

$$H := \begin{pmatrix} 1 & 0 & 1 & 0 & 1 & 0 & 1 \\ 0 & 1 & 1 & 0 & 0 & 1 & 1 \\ 0 & 0 & 0 & 1 & 1 & 1 & 1 \end{pmatrix}.$$

Maximum Likelihood (ML) Decoding

We assume that the channel through which the codewords are sent is *memoryless*, i.e., that the noise affects each individual bit independently.

Definition. A binary-input memoryless channel is characterized by an output domain $\mathcal{Y}$ and the two conditional probability functions

$$P(y_i|x_i = 0) \text{ and } P(y_i|x_i = 1),$$

which specify how the output $y_i \in \mathcal{Y}$ depends on the possible inputs 0 and 1.

A more frequently used concept (we shall need it too) is the *log-likelihood ratio* (LLR),

$$\lambda_i = \ln\left(\frac{P(y_i|x_i = 0)}{P(y_i|x_i = 1)}\right),$$

which represents the entire information revealed by the channel about the sent symbol $\mathbf{x}$. If $\lambda_i > 0$, then $x_i = 0$ is more likely than $x_i = 1$, and vice versa if $\lambda_i < 0$.

Thus, the receiver observes n λ_i values as a result of channel transmission. The decoder has to answer the following problem: *which codeword* $\mathbf{x} \in \mathcal{C}$ *was sent under consideration of* $\boldsymbol{\lambda}$ *(and* $\mathbf{y}$*)?* We speak of decoding success if $\mathbf{x}$ was sent and the decoder decoded the observed $\boldsymbol{\lambda}$ into $\mathbf{x}$, and of a decoding error if the decoded word is not equal to $\mathbf{x}$.

An *optimal* decoder would always return the codeword that was sent with highest probability among all codewords $\mathbf{x} \in \mathcal{C}$, given the observed channel output $\mathbf{y}$:

$$\mathbf{x}_{MAP} = \arg\max_{\mathbf{x} \in \mathcal{C}} P(\mathbf{x}|\mathbf{y}).$$

This is called MAP decoding. It is equivalent to ML decoding:

$$\mathbf{x}_{ML} = \arg\max_{\mathbf{x} \in \mathcal{C}} P(\mathbf{y}|\mathbf{x}),$$

i.e., for codewords with length equal to n, the above is rewritten

$$\mathbf{x}_{ML} = \arg\max_{\mathbf{x} \in \mathcal{C}} \prod_{i=1}^{n} P(y_i|x_i). \tag{2}$$

By simple algebra this is equivalent to

$$\mathbf{x}_{ML} = \arg\min_{\mathbf{x} \in \mathcal{C}} \sum_{i:x_i=1} \ln\left(\frac{P(y_i|0)}{P(y_i|1)}\right),$$

i.e., we have

$$\mathbf{x}_{ML} = \arg\min_{\mathbf{x}\in\mathcal{C}} \sum_{i=1}^{n} \lambda_i x_i, \tag{3}$$

where λ_i is the LLR defined earlier. Therefore, ML decoding is equivalent to minimizing the linear function $\boldsymbol{\lambda}^T\mathbf{x}$ over $\mathbf{x}\in\mathcal{C}$.

Rewrite the problem $\min_{\mathbf{x}\in\mathcal{C}}\boldsymbol{\lambda}^T\mathbf{x}$ as an integer programming problem using the characterization of $\mathcal{C}$ in (1).

1.33 ▪ Truss Topology Design by Mixed Integer Optimization

In truss optimization we want to design a pin-jointed framework called a *truss*. The truss consists of m slender bars of constant mechanical properties characterized by their Young's modulus E. We will consider trusses in a d-dimensional space, where $d = 2$ or $d = 3$.

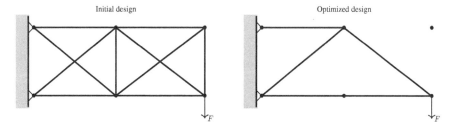

Figure 3. *A Typical 2-Dimensional Truss at Its Initial Configuration and after Optimization.*

The bars are jointed at $\tilde{n}$ nodes. The system is under load, i.e., forces $f_j \in \mathbb{R}^d$ are acting at some nodes j. They are aggregated in a vector $\mathbf{f}$, where we put $f_j = 0$ for nodes j that are not under load. This external load is transmitted along the bars causing displacements of the nodes that make up the displacement vector $\mathbf{u}$. Let p be the number of fixed nodal coordinates, i.e., the number of components with prescribed discrete homogeneous Dirichlet boundary condition. We omit these fixed components from the problem formulation, thus reducing the dimension of u to $n = d \cdot \tilde{n} - p$. Analogously, the external load $\mathbf{f}$ is considered as a vector in $\mathbb{R}^n$. The design variables in the system are the bar cross-sectional areas ("thicknesses") $\{t_1, \dots, t_m\}$. Typically, we want to minimize the weight of the truss. We assume to have a unique material (and thus density) for all bars, so this is equivalent to minimizing the volume of the truss, i.e.,

$$\sum_{i=1}^{m} \ell_i t_i,$$

where ℓ_i is the length of the ith bar. The optimal truss should satisfy mechanical equilibrium conditions:

$$K(\mathbf{t})\mathbf{u} = \mathbf{f};$$

here

$$K(\mathbf{t}) := \sum_{i=1}^{m} t_i K_i, \quad Ki = \frac{E_i}{\ell_i} \gamma_i \gamma_i^T$$

is the so-called stiffness matrix, E_i the Young modulus of the ith bar, and γ_i the n-vector of direction cosines. We further introduce the compliance of the truss $\mathbf{f}^T \mathbf{u}$ that indirectly measures the stiffness of the structure under the force $\mathbf{f}$. The minimum volume truss topology optimization problem under a maximum allowable stiffness of the truss $\bar{\gamma}$ (given) reads as

$$
\begin{aligned}
\text{minimize} \quad & \sum_{i=1}^{m} \ell_i t_i \\
\text{subject to} \quad & K(\mathbf{t})\mathbf{u} = \mathbf{f}, \\
& \mathbf{f}^T \mathbf{u} \leq \bar{\gamma}, \\
& 0 \leq t_i \leq \bar{t}, \quad \forall i \in \{1, \ldots, m\}, \\
& \mathbf{t} \in \mathbb{R}^m, \\
& \mathbf{u} \in \mathbb{R}^n.
\end{aligned}
\qquad (P)
$$

This is a non-linear optimization problem due to the constraint $K(\mathbf{t})\mathbf{u} = \mathbf{f}$.

The optimal truss (see Figure 3), as a solution of the above problem, includes bars that may vary from tiny ones to those on the upper bound $\bar{t}$. Such a result is highly impractical. Hence a natural requirement arises: the bar thicknesses should only be chosen from a small set of discrete values. For the sake of simplicity, assume that these values are integers. So we want to add another constraint to our problem:

$$t_i \in \{0, 1, \ldots, T\}, \quad \forall i \in \{1, 2, \ldots, m\},$$

where T is equal either to 1 (giving a binary variable) or to some small integer, say 3, 4, or 5. Add this constraint to the above formulation to obtain a mixed integer non-linear problem formulation.

1.34 ▪ Sequence Alignment in Computational Biology

In computational biology, one must compare objects which consist of a set of elements arranged in a linearly ordered structure. A typical example is the genomic sequence, which is equivalent to a string over an alphabet P (the 4-letter nucleotide alphabet $\{A, T, C, G\}$, or the 20-letter amino acid alphabet). Aligning two (or more) such objects consists in determining subsets of corresponding elements in each. The correspondence must be order-preserving, i.e., if the ith element of object 1 corresponds to the kth element of object 2, no element following i in object 1 can correspond to an element preceding k in object 2. The situation can be described graphically as follows. Given two objects, where the first has n elements, numbered $\{1, \ldots, n\}$, and the second has m elements, numbered $\{1, \ldots, m\}$, we consider the complete bipartite graph

$$W_{nm} := ([n], [m], L),$$

where $L = [n] \times [m]$ and $[k] := \{1, \ldots, k\}$, $k \in \mathbb{Z}_+$. We call a pair (i, j) with $i \in [n]$ and $j \in [m]$ a *line* (since it aligns i with j), and we denote it by $[i, j]$. L is the set of all lines. Two lines $[i, j]$ and $[i', j']$, with $[i, j] \neq [i', j']$, are said to *cross* if either $i' \geq i$ and $j' < j$ or $i' \leq i$ and $j' > j$ (by definition, a line does not cross itself). Graphically, crossing lines correspond to lines that intersect in a single point. A matching is a subset of lines, no two of which share an endpoint. An alignment is identified by a non-crossing matching, i.e., a matching for which no two lines cross, in W_{nm}.

Define G_L (the *line conflict graph*) as follows. Each line $l \in L$ is a vertex of G_L, and two vertices l and h are connected by an edge if the lines l and h cross.

We define binary variables x_l to select which lines $l \in L$ define the alignment, and constraints to ensure that the selected lines are non-crossing. Show that the following family of clique inequalities can be used as constraints for the alignment problem:

$$\sum_{l \in Q} x_l \leq 1, \quad \forall Q \in \text{clique}(L),$$

where $\text{clique}(L)$ denotes the set of all cliques in G_L.

1.35 ▪ Election Rigging or Gerrymandering

In HeartbreakTown, the party of the politician Evita Macaroni has finally beaten her rivals and won the mayoral elections. However, Suzanne Neckpain from an opposition party is gathering momentum as a credible opponent, and Macaroni wishes to consolidate her position in the city, the M counties of which need to be grouped to electoral districts. For each county i the forecast number of favorable votes F_i for Macaroni and the total number of electors per quarter E_i are known. All electors must vote and the winner needs to have the absolute majority. A valid electoral district is formed by several adjacent quarters and must have between L and U voters. Electoral districts consisting of a single quarter are permitted if it has at least P voters. Determine a partitioning into q electoral districts that maximizes the number of seats for Macaroni. Given the set of all possible electoral districts, we have to choose a subset such that every county appears in a single chosen electoral district.

Based on the given forecasts of favorable votes per county, one can calculate the majority indicator P_d for every district d. If the sum of favorable votes is at least half of the electorate of all counties in this district (that is, Macaroni has the absolute majority), then P_d is equal to 1, and 0 otherwise.

In the following we shall assume that the set R of possible districts has been saved in the form of an array Q. An entry Q_{di} of this array has the value 1 if county i is contained in district d and 0 otherwise. Formulate an integer programming problem which will form a set of districts maximizing the number of seats that will go to Macaroni's party.

1.36 ▪ Traveling Salesperson with Due Dates

Given a directed graph $G = (N, A)$, positive travel time t_{ij} for arc $(i, j) \subset A$, and a due date d_i for each city $i \in N$, a salesperson who is currently at city 1 has to visit each of the remaining cities exactly once no later than its corresponding due date and come back to city 1. We assume there is no due date for returning to city 1. The aim is to find a feasible Traveling Salesperson Problem (TSP) tour respecting the due dates with minimum total travel time. Adapt the Miller–Tucker–Zemlin formulation to model this problem. **Hint:** Define u_i as the arrival time to city i.

1.37 ▪ Prize Collecting Salesperson

Consider a version of the TSP in which the salesperson collects a prize w_k in every city k that he/she visits and pays a penalty c_l to every city l that he/she fails to visit. Suppose that the cost of traveling from city i to city j is t_{ij} and there is a lower bound W on the amount of prize to be collected. Formulate the problem of finding a tour that minimizes the sum of travel costs and penalties of the salesperson.

1.38 ▪ Multiple Salespersons

Suppose that there are m salespersons located at a depot and c_{ij} is the cost of traveling from city i to city j. There are $n > m$ cities to be visited. Formulate the problem of finding a tour for each salesperson such that the total tour cost is minimized and that each city is visited exactly once by only one salesperson.

1.39 ▪ Clustering in Multivariate Data Analysis

Cluster analysis is a significant issue in multivariate data analysis, as it involves partitioning a set of entities into a specific number of groups in an optimal manner. This problem arises in numerous fields.

Let $N = \{1, 2, \ldots, n\}$ be the set of objects or elements that are to be clustered into m clusters. We are given for each $i \in N$, a vector $(x_{1i}, \ldots, x_{pi}) \in \mathbb{R}^p$. Let n^k denote the number of elements in the kth cluster, $k \in \{1, 2, \ldots, m\}$. So we have

$$\sum_{k=1}^{m} n^k = n.$$

Let m_0 denote a given upper bound on n^k. If there is no such restriction, $m_0 = n$. Even though we wish to have only m clusters, we define n clusters fictitiously, $n - m$ of which will have no elements at all. Further, for each nonempty cluster we have a median. Thus there should be m medians in all. We call the cluster for which element j is the median the j-cluster. We define the binary variable x_{ij} as equal to one if ith element belongs to the j-cluster and zero otherwise. Given a distance matrix with elements d_{ij} for every pair i, j of elements, complete the integer programming formulation to find the m clusters so as to minimize total distance.

1.40 ▪ Support Vector Machines: Feature Selection Budget

Consider a training set Ω partitioned into two classes; each object $i \in \Omega$ is represented with a pair $(x_i, y_i) \in \mathbb{R}^n \times \{-1, 1\}$, where n is the number of features analyzed over each element of Ω, x_i contains the features' values, and y_i provides the labels, 1 or -1, associated with the two classes in Ω. The Support Vector Machine (SVM) determines a hyperplane $f(\mathbf{x}) = \mathbf{w}^T \mathbf{x} + \mathbf{b}$ that optimally separates the training examples (see Figure 4). In the case of linearly separable data, this hyperplane maximizes the margin between the two data classes, i.e., it maximizes the distances between two parallel hyperplanes supporting some elements of the two classes. Even if the training data is non-linearly separable, the constructed hyperplane also minimizes classification errors.

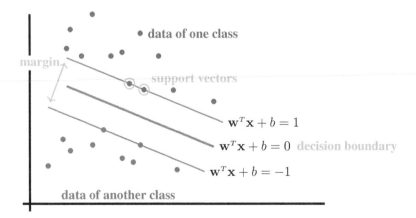

Figure 4. *A 2-Dimensional (e.g., Two Features) Example of Classification.*

Thus, the classical SVM model minimizes an objective function that is a compromise between the structural risk, given by the inverse of the margin, $\|\mathbf{w}\|_2$, and the empirical risk, given by the deviation of misclassified objects. Several SVM models have been proposed using different measures of margin and deviation. Among them, the standard 2-SVM (Bradley and Mangasarian [4]) uses the following quadratic optimization formulation:

$$
\begin{aligned}
\text{minimize} \quad & \tfrac{1}{2}\|\mathbf{w}\|_2^2 + C\sum_{i=1}^m \xi_i \\
\text{subject to} \quad & y_i(\mathbf{w}^T\mathbf{x}_i + b) \geq 1 - \xi_i, \quad \forall i \in \{1,\dots,m\}, \\
& \boldsymbol{\xi} \in \mathbb{R}_+^m, \\
& \mathbf{w} \in \mathbb{R}^n.
\end{aligned}
$$

In Figure 5, slack variables ξ_i, $i \in \{1,\dots,m\}$, measure the deviations of misclassified elements. Additionally, a penalty parameter C that regulates the trade-off between structural and empirical risk is added. The constraints determine whether or not the training data are separable by the classifier hyperplane.

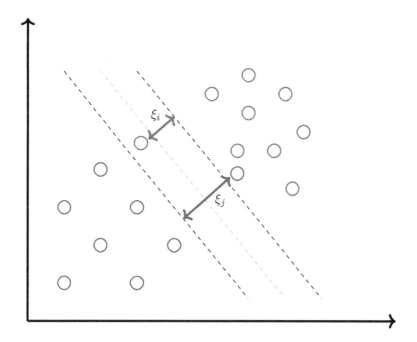

Figure 5. *A 2-Dimensional (e.g., Two Features) Example of Misclassification.*

Structural risk can also be measured in the ℓ_1 norm of $\mathbf{w}$ resulting in the model

$$
\min \|\mathbf{w}\|_1 + C\sum_{i=1}^m \xi_i
$$

subject to identical constraints. An equivalent linear optimization model is thus obtained by introducing auxiliary variables z_j that are equal to the absolute value of w_j at the optimum:

$$
\begin{aligned}
\text{minimize} \quad & \sum_{j=1}^{n} z_j + C \sum_{i=1}^{m} \xi_i \\
\text{subject to} \quad & y_i(\mathbf{w}^T \mathbf{x}_i + b) \geq 1 - \xi_i, \quad \forall i \in \{1, \dots, m\}, \\
& -z_j \leq w_j \leq z_j, \qquad\qquad \forall j \in \{1, \dots, n\}, \\
& \boldsymbol{\xi} \in \mathbb{R}_+^m, \\
& \mathbf{w}, \mathbf{z} \in \mathbb{R}^n.
\end{aligned}
$$

Usually, in practice real data are composed of a few sample elements (m), but each element has a large number of related features (n). Therefore, it is important to select a suitable set of features to construct the classifier. The purpose of this problem is to construct a mixed integer programming model incorporating feature selection into structural and empirical risk minimization. Assume that a cost c_j is associated with including feature j into the classifier, and a total budget B is allotted. Give a mixed integer programming model that minimizes the empirical risk of misclassification under the usual constraints and the budget constraint for feature selection.

1.41 ▪ Network Revenue Management

Lumumba Airlines operates a flight from Congo-Brazzaville to Kampala, and a flight from Kampala to Windhoek. Passengers willing to go from Congo-Brazzaville to Windhoek have to change flights at Kampala. Both airplanes have a capacity equal to 100 seats. The following table gives the relevant information for fare classes, mean demands, and the fares in francs:

OD pair	Fare class	Fare	Mean demand
CB-K 1	Y	500	10
CB-K 1	M	300	20
CB-K 1	Q	150	40
K-W 2	Y	600	15
K-W 2	M	350	25
K-W 2	Q	200	45
CB-W 3	Y	1000	8
CB-W 3	M	625	25
CB-W 3	Q	250	45

(a) Give a linear integer programming formulation to optimize this simple network where the goal is to reserve different fare class seats for passengers in each origin destination pair so as to maximize the total fare collected. The seats reserved for a fare class and origin-destination city pair cannot exceed the mean demand for that origin-destination pair and fare class.

(b) Given that at an optimal solution for CB-K 1 Q class three passengers are refused a seat, for K-W 2 Q class 18 passengers are refused a seat, and for CB-W 3 Q no passenger is given a seat, find the optimal allocations.

1.42 ▪ Shirley's Birthday

Shirley has invited 100 of her friends for her birthday party. Each of them brought a number of coins Shirley may want for a birthday gift. More precisely, each of Shirley's friends brought a different number of silver, gold, and platinum coins. Shirley will pick a coin type from each of her friends, and she wants to maximize her gain. Since she has no materialistic interest in coins, she just wants to maximize the number of coins she can get.

(a) Assume coin types $\{1, 2, 3\}$ correspond to silver, gold, and platinum, respectively. Let c_{ij} be the number of coin type j brought by ith friend for $i \in \{1, \ldots, 100\}$ and $j \in \{1, 2, 3\}$. Formulate an IP model for Shirley to make the optimum decision.

(b) Shirley changes her mind on the decision rule. Now she wants to pick platinum coins from at least 40 of her friends and pick silver coins from no more than 30 of her friends. How would your model change to incorporate this constraint?

Chapter 2

Shortest Paths, Maximal Flows, and Trees

The problems collected in this chapter are connected to some combinatorial problems admitting fast (polynomial time) algorithms and/or exact linear programming (LP) relaxations, e.g., shortest paths, optimum trees, and maximal matchings on various graphs (see Figure 6) with the exception of Steiner trees, which could be covered in other parts of the book. An authoritative and comprehensive source for the topics of this chapter is [1], while [21] is a more introductory and programming oriented source.

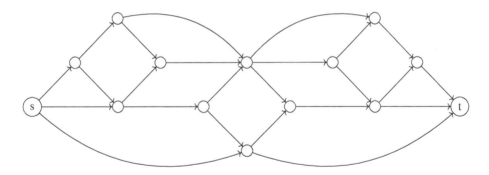

Figure 6. *Network with Labeled Source and Sink Nodes.*

2.1 ▪ Shortest Paths with Two Sources and Two Sinks

Suppose you have the graph in Figure 7 with nodes $\{S_1, S_2, 1, 2, 3, 4, 5, T_1, T_2\}$ and arcs with lengths written on it. Give an integer linear programming model to find the shortest path starting from S_1 or S_2 and reaching T_1 or T_2.

Hint: Define some extra nodes in order to turn this problem into a regular shortest path problem. Then you may use an algorithm that you learned in class.

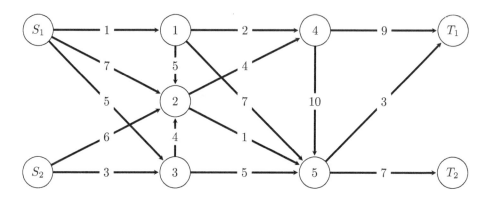

Figure 7. *Network for Problem 2.1.*

2.2 ▪ Shortest Path with Restrictions

Let $G = (V, E)$ be a directed network with a specified source node s in V, a specified destination node t, and cost vector **c**.

(a) Give an integer linear programming formulation to find a path from s to t in such a way that the path uses the minimum number of arcs and the total cost of the path does not exceed a given threshold value M.

(b) How would you formulate the problem of finding the shortest path if the number of nodes visited on a shortest path should not exceed p?

2.3 ▪ Shortest Path Duality

Assume we have the directed graph $G = (V, E)$ (see Figure 8).

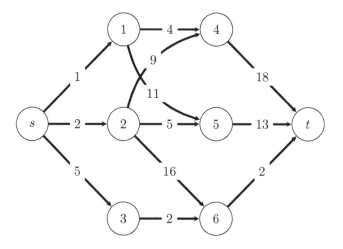

Figure 8. *Network for Problem 2.3.*

(a) Find the shortest path between s and t using Dijkstra's algorithm.

(b) Suppose you have 8 balls (of negligible size), some of which are tied together with inelastic ropes having lengths d_{ij} as you see in graph G. Assume you are holding this chain from ball "t" and standing at a point sufficiently tall such that ball "s" does not touch the ground. Clearly, while some ropes become tight, some are going to stay loose. Only using the variables u_i, which define the height of each ball from the ground, formulate a linear program to find the height difference between balls "t" and "s."

(c) Find the optimal value for part (b).

2.4 ▪ Constrained Shortest Path and Saving Corporal Raymond

Corporal Raymond is trapped in trenches far from the command line. It is possible to reach the trench post where he is stranded by going through a number of safe posts. The distances between those posts are given, and the least distance from the command line to the corporal is to be used because of the enemy fire. A team of volunteers will undertake the operation. Sergeant Pierre requires the team to visit at least half of the posts on the way to the corporal and supply those with ammunition. Formulate this problem as an integer optimization problem.

2.5 ▪ LP Relaxation for Matching

Consider the maximum cardinality matching problem on the undirected graph given in Figure 9.

(a) Find a provably optimal solution to the LP relaxation of the problem.

(b) What is the matching number of the graph?

(c) Add an inequality valid for all matchings that will eliminate the LP relaxation optimum found in (a).

2.6 ▪ LP Relaxation for Edge Covering

Consider the minimal cardinality edge covering problem on the undirected graph given in Figure 9.

(a) Find a provably optimal solution to the LP relaxation of the problem.

(b) What is the covering number of the graph?

Hint: Use LP duality connection to maximal cardinality matching.

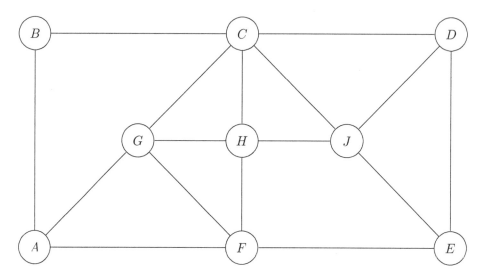

Figure 9. *Network for Problem 2.5.*

2.7 ▪ Pairing Surgeons

An allied forces military field hospital in World War II has many surgeons speaking different languages and with different specialties and previous experience. The hospital administration has to form surgeon teams of two ("crews") for every operation with a compatible language and sufficiently good knowledge of the medical operation to be performed. In the example there are eight surgeons. In the table below every surgeon is characterized by a mark between 0 (worst) and 20 (best) for his language skills (in French, Dutch, English, and Spanish) and for his experience/specialization with different kinds of surgery.

	Surgeon	1	2	3	4	5	6	7	8
Language									
	English	20	15	0	12	0	0	9	8
	French	12	0	0	10	15	19	8	8
	Dutch	0	19	13	0	9	12	13	11
	Spanish	0	0	0	0	17	0	0	17
Specialization/experience									
	General	19	12	16	0	0	0	9	0
	Urology	10	0	9	14	15	9	13	12
	Gastrology	0	16	0	11	12	10	0	0
	Oto-rhino	0	0	15	0	0	12	15	0
	Ophtalmology	0	0	0	0	13	19	0	19

A valid surgeon pair consists of two surgeons that both have each at least $\frac{11}{20}$ for the same language and $\frac{11}{20}$ on the same specialty. Form a suitable graph. Give a formulation that will check whether it is possible to assign all surgeons.

2.8 ▪ Steiner Trees

Given is a directed graph $G = (V, E)$ with arc weights c_{ij}, and let $r \in V$ be a given root node. Also given is a subset of *destination* nodes $T \subseteq V \setminus \{r\}$. The Steiner Tree problem consists in identifying a partial arborescence (an arborescence is a directed version of a tree graph, connected, and acyclic) rooted at r and allowing it to reach every vertex in T at minimal cost. The Steiner tree may also touch one or more vertices in $V \setminus (T \cup \{r\})$. A typical application of this problem is in telecommunications where we wish to send a signal to some receivers passing by intermediate nodes if necessary. (See Figure 10.)

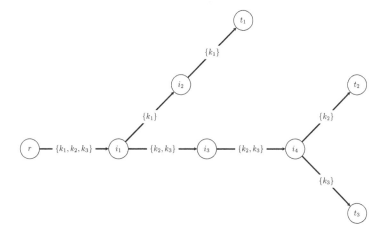

Figure 10. *A Steiner Tree Example: t_1, t_2, t_3 Are the Destination Nodes.*

Formulate the Steiner Tree problem as a integer optimization problem.

2.9 ▪ Maximum Weight Spanning Trees

A tree is a graph which is connected and has no cycles. A spanning tree of an undirected graph $G = (V, E)$ is a tree subgraph of G containing all nodes of the graph G. Assuming all edges $e \in E$ have positive weights c_e, we are interested in finding a maximum total weight spanning tree of G. Formulate an integer optimization problem for this purpose.

Hint: A spanning tree of G has exactly $|V| - 1$ edges.

2.10 ▪ Minimum Weight Spanning Trees

With reference to the previous problem, assuming all edges $e \in E$ have positive weights c_e, we are interested in finding a minimum total weight spanning tree of G. Formulate an integer optimization problem for this purpose.

2.11 ▪ Disasters and Spanning Trees

A town has been hit by a devastating natural disaster, leaving many houses in need of urgent assistance. The town has n houses, and the roads between them may be partially blocked due to the disaster's aftermath. The probability of encountering roadblocks or obstructions along these routes is denoted as p_{ij} for each pair of houses i and j.

Your mission is to organize a disaster relief operation to serve all the houses while minimizing the overall risk of encountering roadblocks and obstructions. To accomplish this, you need to plan a network of alternative routes, like cleared paths or temporary bridges, connecting all the houses and ensuring the lowest total probability of encountering roadblocks or obstacles while delivering aid.

In this context, the problem can be reformulated as finding the minimum or maximum weight spanning tree within the network of houses and available routes, where the edge weights represent the probability of facing roadblocks or obstructions. Your objective is to establish the most efficient and reliable disaster relief distribution network within this affected town, ensuring aid reaches all the houses while minimizing the risk associated with blocked roads.

Hint: Assume independence in the sense of probability theory and use natural logarithms.

2.12 ▪ Emergency Planning and Spanning Trees I

The (bidirectional) road connections among towns in Fairway County will be designated for emergency evacuation in case of a natural disaster. The towns have been partitioned into two subsets of towns E_1 and E_2. For every road connection e between towns i and j, there is a transport cost c_e and an importance coefficient w_e that have been calculated by engineers. We wish to find a minimum cost connection among towns satisfying the following requirements:

- the total weight of selected road connections should not be less than a given W;

- there should not be cycles in the emergency road network formed;

- the number of selected roads chosen from E_1 should be less than the number of those selected from E_2.

Give an integer programming formulation for the above problem.

2.13 ▪ Emergency Planning and Spanning Trees II

Continuing the scenario of the previous exercise, we wish to find a minimum cost connection among towns satisfying the following requirements:

- there should not be cycles in the emergency road network formed;

- the average weight of selected road connections from E_1 should not exceed a given W;

- at least of half of the road connections chosen chosen should come from E_1.

Give an integer programming formulation for the above problem.

2.14 ▪ Envy-Free Perfect Matching

An internet site allows you to sell your rare classical music CD collection online. Your collection has n CDs. The selling procedure works as follows: every potential buyer $b \in \{1, \ldots, n\}$ (the site allows exactly n people to participate) posts online his/her valuations v_{bi} for every CD $i \in \{1, \ldots, n\}$ (e.g., $v_{13} = 10$ means customer 1 values CD 3 at 10 dollars). Having collected the valuations matrix V, the site decides which customer gets which CD, and what price he/she has to pay. The rule is that, the CDs being rare collection items, every customer can only get one CD. The site is trying to maximize its total revenue. However, the following rules have to be observed:

- If a customer b is assigned a CD i at a price p_i then this must be one of the most advantageous CDs for him/her in terms of utility, i.e.,

$$v_{bi} - p_i \geq v_{b\ell} - p_\ell, \quad \forall \ell \neq i.$$

- If a customer b is assigned a CD i at price p_i, then his/her utility must be non-negative. i.e., $v_{bi} - p_i \geq 0$.

Give a linear integer programming model for the internet site's problem of deciding which customers get CDs and at what price.

2.15 ▪ Spies and B-Matchings

Agents stationed over a geographical area in enemy territory are to communicate with one another over secure channels. Establishing a direct communication link e between agents i and j incurs a cost c_e to the intelligence office due to stringent security exigencies. Each agent v should establish a communication link with exactly b_v other agents. One wishes to find a minimum total cost connection pattern satisfying the above requirement. Formulate the problem as an integer optimization problem.

2.16 ▪ Matchings and Stable Marriages

In the stable marriage problem there are two sets of agents, the set of men $M = \{m_1, \ldots, m_p\}$, and the set of women $W = \{w_1, \ldots, w_q\}$. Each agent has a complete, transitive, and strict preference ordering over the agents of the other side, and the prospect of remaining single. We say that a pair $(m, w) \in M \times W$ is acceptable if m and w prefer each other to remaining single. The set of acceptable pairs is denoted A.

A matching μ (of men to women) is called *individually rational* if no agent a prefers being single to being married to $\mu(a)$. A matching is *stable* if it is individually rational and there is no pair $(m, w) \in A$ such that both man m prefers woman w to $\mu(m)$ and woman w prefers man m to $\mu(w)$.

We write $a >_c b$ if person c prefers person a to person b, and $a \geq_c b$ if $a = b$ or $a >_c b$ (i.e., person c is indifferent between a and b or if she/he prefers a to b).

Write a set of linear inequalities that characterize stable and individually rational matchings.

2.17 ▪ Stable Marriage with Ties and Incomplete Lists

In this problem, we shall consider an extension of the stable marriage problem in the context of adoptions.

An instance of the Stable Marriage problem with Ties and Incomplete lists (SMTI) comprises a set C of n_1 children and a set F of n_2 families, where each child (respectively, family) ranks a subset of the families (respectively, children) in order of preference, possibly with ties. We say that a child $c \in C$ finds a family $f \in F$ *acceptable* if f belongs to c's preference list, and we define acceptability for a family in a similar way. We assume that preference lists are *consistent*, that is, given a child-family pair $(c, f) \in C \times F$, c finds f acceptable if and only if f finds c acceptable. If c does find f acceptable, then we call (c, f) an acceptable pair.

A matching M is a subset of acceptable pairs such that, for each agent $a \in C \cup F$, a appears in at most one pair in M. If a appears in a pair of M, we say that a is matched; otherwise a is unmatched. In the former case, $M(a)$ denotes a's partner in M, that is, if $(c, f) \in M$, then $M(c) = f$ and $M(f) = c$.

Let M be a matching. A child-family pair $(c, f) \in (C \times F) \setminus M$ is called a *blocking pair* of M, or *blocks* M, if

- (c, f) is an acceptable pair,

- either c is unmatched in M or c prefers f to $M(c)$, and

- either f is unmatched in M or f prefers c to $M(f)$.

Stability of a matching is defined next. M is said to be *stable* if it admits no blocking pair. In SMTI, the goal is to find an arbitrary stable matching.

The purpose of this problem is to represent SMTI as the 0-1 solutions of a linear system of equations/inequalities.

Let us use i and j to represent a child and family, rather than c and f, respectively, as i and j are by convention more typically used as subscript variables. Consider the following notation:

- $F(i)$ is the set of families acceptable for child $i \in \{1, \dots, n_1\}$.

- $C(j)$ is the set of children acceptable for family $j \in \{1, \dots, n_2\}$.

- $r_j^c(i)$ is the rank of family j for child i, defined as the integer k such that j belongs to the kth most-preferred tie in i's list ($i \in \{1, \dots, n_1\}$, $j \in F(i)$). The smaller the value of $r_j^c(i)$, the better family j is ranked for child i.

- $r_i^f(j)$ is the rank of child i for family j, defined as the integer k such that i belongs to the kth most-preferred tie in j's list ($j \in \{1, \dots, n_2\}$, $i \in C(j)$). The smaller the value of $r_i^f(j)$, the better child i is ranked for family j.

- $F_j^{\leq}(i)$ is the set of families that child i ranks at the same level or better than family j, that is, $F_j^{\leq}(i) = \{j' \in F : r_{j'}^c(i) \leq r_j^c(i)\}$, ($i \in \{1, \dots, n_1\}, j \in F(i)$).

- $C_i^{\leq}(j)$ is the set of children that family j ranks at the same level or better than child i, that is, $C_i^{\leq}(j) = \{i' \in C : r_{i'}^f(j) \leq r_i^f(j)\}$ ($j \in \{1, \dots, n_2\}; i \in C(j)$).

Using the above, give a formulation such that every binary solution is an SMTI.

2.18 ▪ Hospitals and Residents with Ties

An instance of the Hospitals/Residents problem with Ties (HRT) comprises a set D of n_1 resident doctors and a set H of n_2 hospitals. Each doctor (respectively, hospital) ranks a subset of the hospitals (respectively, doctors) in order of preference, possibly with ties. Additionally, each hospital h has a capacity $c_h \in \mathbb{Z}_+$, meaning that h can be assigned at most c_h doctors, while each doctor is assigned to at most one hospital. The definitions of the terms "consistent" and "acceptable" are analogous to the SMTI case.

A matching M is a subset of acceptable pairs such that each doctor appears in at most one pair, and each hospital $h \in H$ appears in at most c_h pairs. Given a doctor $d \in D$, the terms "matched" and "unmatched," and the notation $M(d)$, are defined as in the SMTI case. Given a hospital $h \in H$, we let $M(h) = \{d \in D : (d, h) \in M\}$. We say that h is full or under-subscribed in M if $|M(h)| = c_h$ or $|M(h)| < c_h$, respectively. We next define stability for the HRT case.

Let M be a matching. A doctor-hospital pair $(d, h) \in (D \times H) \setminus M$ is a blocking pair of M, or blocks M, if

- (d, h) is an acceptable pair,

- either d is unmatched in M or d prefers h to $M(d)$, and

- either h is under-subscribed in M or h prefers d to some member of $M(h)$.

M is said to be *stable* if it admits no blocking pair.

The mathematical formulation uses the same notation that was used for SMTI in the previous problem except that

- the term "family" is replaced by "hospital" and $F(i), r_i^f(j)$, and $F_{\bar{j}}^{\leq}(i)$ are changed into $H(i), r_i^h(j)$, and $H_{\bar{j}}^{\leq}(i)$, respectively;

- the term "child" is replaced by "doctor" and $C(j), r_j^c(i)$, and $C_{\bar{i}}^{\leq}(j)$ are changed to $D(j), r_j^d(i)$, and $D_{\bar{i}}^{\leq}(j)$, respectively;

- the capacity of hospital $h \in \{1, \ldots, n_2\}$ is denoted by c_h.

Using the above, give a formulation such that every binary solution is a stable matching for HRT.

2.19 ▪ Maximum Number of Disjoint Paths in a Graph

Let $G = (V, E)$ be a directed graph with two specified nodes s, the source, and t, the sink. You want to find out if there exist at least two directed paths from s to t which do not share any arc (i.e, they do not have any arc in common).

(a) Formulate this problem as a maximum flow problem.

(b) Based on your answer to part (a), what do you conclude using maximum flow-minimum cut duality?

2.20 ▪ Matrix Rounding and Maximum Flow

The matrix rounding problem deals with a $p \times q$ matrix, denoted as D, consisting of real numbers. Each row in this matrix has a known sum a_i, and each column has a known sum b_j. The goal is to round each element d_{ij} down or up, as well as the row sums a_i and column sums b_j, to the nearest integer in a way that maintains the row and column sums.

The US Census Bureau used this technique to avoid identification of specific persons from the census data. Consider the following example below: Table 1 represents the original matrix, and Table 2 shows the matrix after rounding.

Table 1. *A Matrix for Rounding.*

4.14	5.82	7.32	17.28
9.6	1.42	1.71	12.73
3.7	2.21	5.51	11.42
17.44	9.45	14.54	

Table 2. *Matrix after Rounding.*

4	6	7	17
9	1	2	12
4	2	5	11
17	9	14	

The objective of this problem is to demonstrate that the task of determining whether a valid integer rounding is possible can be transformed into a maximum flow problem on a well-defined graph.

Chapter 3

Studying Integer Optimization Problems

In this chapter, problems requiring the use of different techniques to deal with the solution and/or structure of integer optimization problems or integer sets are presented. Some problems serve to introduce more advanced topics and/or rely on knowledge presented in integer optimization texts such as [26], which is a good source from which to study integer optimization. Two other and more comprehensive (and somewhat harder to read) textbooks are [15] and [5], which are rather aimed at graduate students. An older reference (most likely out of print) is [19], where some topics such as Lagrangean relaxation and subgradient methods are generously discussed.

3.1 ▪ Integer Points in 2D

Suppose we have the following region (see Figure 11):

$$S = \{(x, y) \in \mathbb{R}_+^2 \mid 3x + 5y \leq 11, \ 7x + 4y \geq 3\}.$$

(a) Plot the region defined by S.

(b) List the elements of set $S \cap \mathbb{Z}^2$ explicitly.

(c) Find the smallest polyhedron $\mathcal{P}$ such that $\mathcal{P} \cap \mathbb{Z}^2 = S \cap \mathbb{Z}^2$.
   Hint: Find a set of inequalities to define $\mathcal{P}$ using your work on parts **(a)** and **(b)**.

3.2 ▪ A Binary Knapsack

Suppose you have the following Binary Knapsack problem:

$$\begin{aligned}
\text{maximize} \quad & 3x_1 + 4x_2 + 5x_3 + 6x_4 \\
\text{subject to} \quad & 2x_1 + x_2 + x_3 + x_4 \leq 3, \qquad (P_1) \\
& \mathbf{x} \in \mathbb{B}^4.
\end{aligned}$$

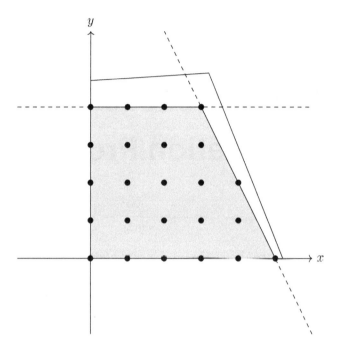

Figure 11. *Convex Hull of the Integer Points in a 2D Polytope.*

(a) Solve the problem using Branch-and-Bound.

(b) Investigate the following integer program:

$$\begin{aligned}
\text{maximize} \quad & 3x_1 + 4x_2 + 5x_3 + 6x_4 \\
\text{subject to} \quad & 97x_1 + 32x_2 + 25x_3 + 20x_4 \leq 134, \\
& \mathbf{x} \in \mathbb{B}^4.
\end{aligned} \qquad (P_2)$$

Show that the two problems have equivalent feasible region and find the optimal solution to (P_2).

3.3 ▪ Hierarchy among LP Relaxations

The following integer sets are given:[1]

$$S_1 := \{\mathbf{x} \in \mathbb{B}^4 : 97x_1 + 32x_2 + 25x_3 + 20x_4 \leq 139\},$$
$$S_2 := \{\mathbf{x} \in \mathbb{B}^4 : 2x_1 + x_2 + x_3 + x_4 \leq 3\},$$
$$S_3 := \{\mathbf{x} \in \mathbb{B}^4 : x_1 + x_2 + x_3 \leq 2, x_1 + x_3 + x_4 \leq 2, x_1 + x_2 + x_4 \leq 2\}.$$

Denote by $LP(S_i)$, $i \in \{1, 2, 3\}$, their LP relaxation sets.

(a) Show that $S_1 = S_2 = S_3$.

---

[1] adapted from a similar problem in [15].

(b) Show that $LP(S_2) \subseteq LP(S_1)$ and $LP(S_3) \subseteq LP(S_1)$. Are the inclusions strict?

(c) What is the relationship between S_2 and S_3?

3.4 ▪ LP Relaxations and Unboundedness

For integer programming problems, LP relaxations are used to find bounds on the optimal value. If an LP relaxation is bounded, then the original problem is bounded. However, an unbounded LP relaxation does not imply unboundedness of the original problem.

(a) Find an example where both the LP relaxation and the original problem are unbounded.

(b) Find an example where LP relaxation is unbounded but the original problem is not.

3.5 ▪ Vertices of an Integral Polyhedron

Suppose you have the following integer optimization problem:

$$\begin{aligned}
\text{maximize} \quad & 3x + 2y \\
\text{subject to} \quad & x^2 + y^2 \leq 50, \\
& x, y \in \mathbb{Z}.
\end{aligned}$$

List the vertices of the smallest polyhedron containing all feasible solutions. What is the optimal solution?

3.6 ▪ Branch-and-Bound Fails

Suppose we have the following optimization problem:

$$\begin{aligned}
\text{minimize} \quad & x_3 \\
\text{subject to} \quad & 12x_1 + 9x_2 - x_3 = 0, \\
& x_3 \geq 1, \\
& \mathbf{x} \in \mathbb{Z}^3.
\end{aligned}$$

Obviously $(0, \frac{1}{9}, 1)^T$ is a linear relaxation solution.

(a) Find an optimal solution using Branch-and-Bound. If you cannot find any, explain why Branch-and-Bound fails.

(b) Add a clever cut, which makes the Branch-and-Bound algorithm functional again. Then find an optimal solution to the problem.

3.7 ▪ Covers and Binary Knapsack

Suppose we have the following Binary Knapsack problem:

$$\begin{aligned}\text{maximize} \quad & x_1 + x_2 + x_3 \\ \text{subject to} \quad & 3x_1 + 4x_2 + 7x_3 + 9x_4 \leq 11, \\ & \mathbf{x} \in \mathbb{B}^4.\end{aligned}$$

(a) Find the LP relaxation solution.

(b) Find a minimal cover and cover inequality to eliminate the LP solution you found in part (a).

(c) Can you improve the inequality you found in part (b)? Explicitly state which fractional point you cut with this improvement.

3.8 ▪ Bin Packing Covers

Consider the Bin Packing problem of Problem 1.2. Let $C \subset \mathcal{N}$ be a subset of the items such that the total size of the items in C is larger than b, that is,

$$\sum_{i \in C} a_i > b.$$

The purpose of this problem is to show that the inequality

$$\sum_{i \in C} x_{ij} \leq (|C| - 1)y_j$$

is a valid inequality for $j \in \{1, \ldots, n\}$ for all feasible solutions of the formulations given in Problem 1.2.

3.9 ▪ Integer Programming Solvers and Accuracy

Suppose you are using an MIP solver with sensitivity 10^{-4}, which implies every x with $|x| < 10^{-4}$ is considered as $\tilde{x} = 0$. A method for logical expressions is the big-M constraints. Assume we try to implement the statement

$$\text{``}x = 0\text{''} \text{ implies } \text{``}y = 0\text{''}$$

where x is binary and y is a non-negative continuous variable bounded above by 10^6. Write the appropriate big-M constraint and explain why it is going to fail computationally.

3.10 ▪ A Quadratic Integer Program

Let $Q \in \mathbb{R}^{n \times n}$, $\mathbf{p} \in \mathbb{R}^n$, and $r \in \mathbb{R}$. Suppose you have the following integer quadratic program:

$$
\begin{aligned}
\text{minimize} \quad & \mathbf{x}^T Q \mathbf{x} + \mathbf{p}^T \mathbf{x} + r \\
\text{subject to} \quad & \frac{i}{n} x_i \le x_{i+1}, \qquad \forall i \in \{1, \dots, n-1\}, \\
& \frac{\sum_{i=1}^{n} x_i}{n} \ge 1 - x_1, \\
& \mathbf{x} \in \mathbb{B}^n.
\end{aligned}
$$

Find the optimal solution by presolving. Can you find a cut constraint to make the problem trivial?

3.11 ▪ Linearizing a Quadratic Integer Program

Let $Q \in \mathbb{R}^{n \times n}$, $\mathbf{c} \in \mathbb{R}^n$. Suppose you have the following integer quadratic program over the variables $\mathbf{x} \in \mathbb{R}^n$:

$$
\begin{aligned}
\text{minimize} \quad & \mathbf{x}^T Q \mathbf{x} + \mathbf{c}^T \mathbf{x} = \sum_{i=1}^{n} \sum_{j=1}^{n} Q_{ij} x_i x_j + \sum_{i=1}^{n} c_i x_i \\
\text{subject to} \quad & \mathbf{x} \in \mathbb{B}^n.
\end{aligned}
$$

Find an equivalent linear integer programming problem.

3.12 ▪ Set Covering

Let $A \in \mathbb{B}^{m \times n}$ and $\mathbf{c} \in \mathbb{R}^n$. An integer linear formulation for the set covering problem (SCP) is

$$
\begin{aligned}
\text{minimize} \quad & \mathbf{c}^T \mathbf{x} \\
\text{subject to} \quad & A \mathbf{x} \ge \mathbf{1}, \qquad\qquad\qquad \text{(SCP)} \\
& \mathbf{x} \in \mathbb{B}^n.
\end{aligned}
$$

Show that the following facts hold:

(a) Let $c_j < 0$. Then $x_j = 1$ for any optimal solution of (SCP).

(b) Consider a vector $\mathbf{e}_j$ of dimension n, which has a value of 1 in the jth position and 0 in all other positions. If there is some $i \in \{1, \dots, m\}$ such that the ith row of A is $\mathbf{e}_j^T$, then $x_j = 1$ for any feasible solution of (SCP).

(c) Let $c_j > c_k \ge 0$ and $\mathbf{a}_k \ge \mathbf{a}_j$, where $\mathbf{a}_k, \mathbf{a}_j$ are the kth and jth columns of A, respectively. Then $x_j = 0$ for any optimal solution of (SCP).

3.13 ▪ Magic Squares Revisited

Let n be a positive integer and S_n be a magic square with magic constant M_n. Then assume we have the following problem:

$$
\begin{aligned}
\text{minimize} \quad & \mathbf{1}^T \mathbf{x} \\
\text{subject to} \quad & S_n \mathbf{x} = M_n \mathbf{1}, \\
& \mathbf{x} \in \mathbb{Z}_+^n.
\end{aligned}
\qquad (P)
$$

Find the optimal solution of (P).

3.14 ▪ Independent or Stable Sets

Let $G = \{V, E\}$ be an undirected graph (with no self-loops) with node set V and edge set E. Let $A \in \mathbb{B}^{|E| \times |V|}$ be the adjacency matrix of G and $\mathbf{a}_i$ be the ith column of A. We define the *Maximum Independent Set* problem as

$$
\begin{aligned}
\text{maximize} \quad & \sum_{i \in V} x_i \\
\text{subject to} \quad & A\mathbf{x} \le \mathbf{1}, \\
& \mathbf{x} \in \mathbb{B}^{|V|}.
\end{aligned}
\qquad (\text{MISP})
$$

Show that the following facts hold:

(a) If there is an $i^* \in V$ such that $\mathbf{1}^T \mathbf{a}_{i^*} = |V| - 1$ and there are $j, k \in V$ such that $(j, k) \notin E$, then $x_{i^*} = 0$ for any optimal solution of (MISP).

(b) If $\frac{|V|^2 - 2|E|}{|V|} \le 1$, then there are $|V|$ alternative optima and a unique non-optimal feasible solution of (MISP).

(c) If $\mathbf{1}^T \mathbf{a}_i = 2$ for any $i \in V$, then the objective is at most $\left\lfloor \frac{|E|}{2} \right\rfloor$ for any feasible solution of (MISP).

3.15 ▪ Chvátal–Gomory Cuts

Consider the following integer programming problem:

$$
\begin{aligned}
\text{maximize} \quad & 2x_1 + x_2 \\
\text{subject to} \quad & 7x_1 + x_2 \le 28, \\
& -x_1 + 3x_2 \le 7, \\
& -8x_1 - 9x_2 \le -32, \\
& \mathbf{x} \in \mathbb{Z}_+^2.
\end{aligned}
\qquad (P_2)
$$

(a) Find a choice of $u_1, u_2, u_3 \ge 0$ that yields the Chvátal–Gomory inequality $x_1 + x_2 \ge 4$.

(b) Is there a choice of $u_1, u_2, u_3 \ge 0$ that yields the valid inequality $x_1 \ge 2$ as a Chvátal–Gomory inequality?

3.16 ▪ Extended Covers

Let $S = \{\mathbf{x} \in \mathbb{B}^n | \sum_{j=1}^{n} a_j x_j \leq b\}$, where a_j and b are positive integers satisfying $a_j \leq b$ for all j. Let $C = \{1, 2, \ldots, k\}$ be a cover and $E(C) = \{1, 2, \ldots, k, k+1, \ldots, m\}$ be the corresponding extended cover (without loss of generality, assume that $a_k \geq a_i$ for all $i \in \{1, 2, \ldots, k\}$).

(a) Write down the extended cover inequality corresponding to the extended cover $E(C)$.

(b) Obtain an inequality as a Chvátal–Gomory inequality that is at least as strong as the extended cover inequality corresponding to the extended cover $E(C)$.

3.17 ▪ Diagonal Distances

Suppose we have a bounded convex set S defined as

$$S = \left\{ (x, y) \in \mathbb{R}^2 | a_i x + b_i y \leq c_i, \quad \forall i \in \{1, \ldots, m\} \right\}$$

for $a, b, c \in \mathbb{R}^m$. Let $\mathcal{S}$ be the set of all squares (including rotations) inside S with *integral corner points*. For any $s \in \mathcal{S}$ we define the *diagonal distance* as

$$d(s) = |x_1 - x_2| + |y_1 - y_2|,$$

where (x_1, y_1) and (x_2, y_2) are the coordinates of two opposing corners of s. Then our goal is to solve the following optimization problem:

$$\max d(s) \text{ such that } s \in \mathcal{S}. \tag{P}$$

Formulate a mixed-integer linear program equivalent to (P).

Hint: Remember that S is a convex set. If you have four corners of a square in S, then the square is automatically included in S.

3.18 ▪ AND Operator

Let $x_1, \ldots, x_n$ and r be binary variables. Suppose $r = \text{AND}(x_1, \ldots, x_n)$. This definition implies the following relations:

(i) $r = 1 \Rightarrow x_i = 1, \forall i \in \{1, \ldots, n\}$,

(ii) $x_i = 1, \forall i \in \{1, \ldots, n\} \Rightarrow r = 1$,

(iii) $\exists i : x_i = 0 \Rightarrow r = 0$,

(iv) $r = 0$ and $x_i = 1, \forall i \in \{1, \ldots, n\} \setminus \{j\} \Rightarrow x_j = 0$.

(a) Find a formulation consisting of $n+1$ constraints to represent the relation between r and x_i; verify that your formulation satisfies all four relations presented above.

(b) Find a formulation consisting of 2 constraints to represent the relation between r and x_i; verify that your formulation satisfies all four relations presented above.

(c) Find a fractional point which is contained in your second formulation but not in the first one, and conclude that the first formulation is stronger.

3.19 ▪ Handling Bounds

Let $\mathbf{a} \in \mathbb{R}^n$, $b \in \mathbb{R}$, and a linear constraint $\mathbf{a}^T\mathbf{x} \le b$ and bounds $\mathbf{l} \le \mathbf{x} \le \mathbf{u}$ be given. We define

$$a_{\min} := \min\{\mathbf{a}^T\mathbf{x} : \mathbf{l} \le \mathbf{x} \le \mathbf{u}\},$$
$$a_{\max} := \max\{\mathbf{a}^T\mathbf{x} : \mathbf{l} \le \mathbf{x} \le \mathbf{u}\}.$$

Show that

(a) $a_{\min} = \sum_{j=1}^{n} \left(l_j \max\{a_j, 0\} + u_j \min\{a_j, 0\} \right).$

(b) $a_{\max} = \sum_{j=1}^{n} \left(u_j \max\{a_j, 0\} + l_j \min\{a_j, 0\} \right).$

(c) If $a_{\min} > b$, then the problem is infeasible.

(d) If $a_{\max} \le b$, the constraint is redundant.

(e) Let $a_i > 0$. If $x \in \mathbb{Z}^n$, then $x_i \le \lfloor \frac{b-a_{\min}+a_i l_i}{a_i} \rfloor$ is a valid inequality.

(f) Let $a_i < 0$. If $x \in \mathbb{Z}^n$, then $x_i \ge \lceil \frac{b-a_{\min}+a_i u_i}{a_i} \rceil$ is a valid inequality.

3.20 ▪ Deep Learning and Neural Networks

Deep learning is proving to be very powerful at dealing with predictive tasks arising in areas such as image classification, speech recognition, machine translation, and robotics and control. The workhorse model in deep learning is the feed-forward network

$$\mathrm{NN} : \mathbb{R}^{m_0} \mapsto \mathbb{R}^{m_s}$$

that maps an input $\mathbf{x}^0 \in \mathbb{R}^{m_0}$ to an output $\mathbf{x}^s = \mathrm{NN}(\mathbf{x}^0) \in \mathbb{R}^{m_s}$. A feed-forward network, an example of which can be seen in Figure 12, with s layers can be recursively described as

$$x_j^i = \mathrm{NL}^{i,j}(w^{i,j} \cdot x^{i-1} + b^{i,j}), \ \forall i \in [[s]], j \in [[m_i]],$$

where $[[n]] = \{1, \ldots, n\}$, m_i is both the number of neurons in layer i and the output dimension of the neurons in layer $i - 1$ (with the input $\mathbf{x}^0 \in \mathbb{R}^{m_0}$ considered to be the 0th layer). Furthermore, for each $i \in [[s]]$ and $j \in [[m_i]]$, $\mathrm{NL}^{i,j} : \mathbb{R} \mapsto \mathbb{R}$ is some simple nonlinear activation function that is fixed before training, and $w^{i,j}$ and $b^{i,j}$ are the weights and bias of an affine function which is learned during the training procedure. In its simplest and most common form, the activation function would be the rectified linear unit (ReLU), defined as $\mathrm{ReLU}(v) = \max\{0, v\}$.

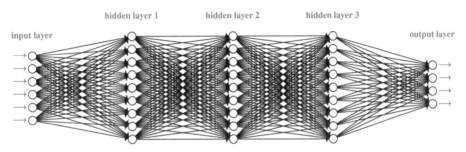

Figure 12. *A Deep Neural Network.*

A standard way to model a function $g : D \subseteq \mathbb{R}^\eta \mapsto \mathbb{R}$ using integer programming is to construct a formulation of its graph defined by $\mathrm{gr}(g; D)$ defined as

$$\mathrm{gr}(g; D) = \{(\mathbf{x}, y) \in \mathbb{R}^\eta \times \mathbb{R} | \mathbf{x} \in D, y = g(\mathbf{x})\}.$$

The focus of this problem is on constructing integer optimization formulations for the graph of individual neurons:

$$\mathrm{gr}(\mathrm{NL} \circ f; D) = \{(\mathbf{x}, y) \in \mathbb{R}^\eta \times \mathbb{R} | \mathbf{x} \in D, y = (\mathrm{NL} \circ f)(\mathbf{x})\}.$$

As an exercise, you are asked to consider the ReLU in the simplest possible setting: where the input is univariate. Take the two-dimensional set $\mathrm{gr}(\mathrm{ReLU}; [l, u])$, where $D = [l, u]$ is some interval in $\mathbb{R}$ containing zero i.e., y is obtained by projecting the input x onto the interval $\mathbb{R}_+$. More precisely, if $x \in \mathbb{R}_+$, then $y = x$; if $x < 0$, then $y = 0$. Construct a mixed-integer formulation for this univariate ReLU.

3.21 ▪ Deep Neural Networks II

A more realistic setting compared to the previous problem is an η-variate ReLU nonlinearity whose input is some η-variate affine function $f : [\mathbf{l}, \mathbf{u}] \mapsto \mathbb{R}$, where $\mathbf{l}, \mathbf{u} \in \mathbb{R}^\eta$ and $[\mathbf{l}, \mathbf{u}] = \{\mathbf{x} \in \mathbb{R}^\eta : l_i \le x_i \le u_i, \forall i \in [[\eta]]\}$ (i.e., the η-variate ReLU nonlinearity given by $\mathrm{ReLU} \circ f$). The box-input $[\mathbf{l}, \mathbf{u}]$ corresponds to known (finite) bounds on each component, which can be estimated depending on the nature of the application.

One can model the graph of the η-variate ReLU neuron as a simple composition of the graph of a univariate ReLU activation function and an η-variate affine function:

$$\text{gr}(\text{ReLU} \circ f; [\mathbf{l}, \mathbf{u}]) = \left\{ (\mathbf{x}, y) \in \mathbb{R}^{\eta+1} \,\middle|\, \begin{array}{c} (f(\mathbf{x}), y) \in \text{gr}(\text{ReLU}; [m^-, m^+]) \\ \mathbf{l} \leq \mathbf{x} \leq \mathbf{u} \end{array} \right\},$$

where $m^- = \min_{\mathbf{l} \leq \mathbf{x} \leq \mathbf{u}} f(\mathbf{x})$ and $m^+ = \max_{\mathbf{l} \leq \mathbf{x} \leq \mathbf{u}} f(\mathbf{x})$.

(a) Using the solution of the previous problem give a mixed-integer representation for

$$\text{gr}(\text{ReLU} \circ f; [\mathbf{l}, \mathbf{u}]).$$

(b) For $f(\mathbf{x}) = x_1 + x_2 - 1.5$ over the domain $[0, 1]^2$, i.e., $\mathbf{l} = [0 \ 0]^T$ and $\mathbf{u} = [1 \ 1]^T$, write down the above mixed integer formulation and its LP relaxation, i.e., for $\text{gr}(\text{ReLU} \circ f; [0, 1]^2)$, and check whether the LP relaxation has points that the convex hull of the integer feasible solutions does not have.

3.22 ▪ Strengthening Covers

Consider the following region for a knapsack problem:

$$S = \{\mathbf{x} \in \mathbb{B}^7 : 11x_1 + 6x_2 + 6x_3 + 5x_4 + 5x_5 + 4x_6 + x_7 \leq 19\}.$$

For S, $C = \{3, 4, 5, 6\}$ is a cover with cover inequality $x_3 + x_4 + x_5 + x_6 \leq 3$.

(a) Suppose we have an inequality c_1, explicitly written as $\alpha_1 x_1 + x_3 + x_4 + x_5 + x_6 \leq 3$. Formulate a mixed integer linear program to find the largest possible α_1 in order to have a valid c_1, and solve it.

(b) Using the α_1 you found in part (a), suppose we have an inequality c_2, explicitly written as $\alpha_1 x_1 + \alpha_2 x_2 + x_3 + x_4 + x_5 + x_6 \leq 3$. Formulate a mixed integer linear program to find the largest possible α_2 in order to have a valid c_2, and solve it.

(c) Find $E(C)$ and write the corresponding extended cover inequality. Denote it by c_3.

(d) Compare the strength of c_2 and c_3.

3.23 ▪ Lifting

Consider the following region for a knapsack problem:

$$S = \{\mathbf{x} \in \mathbb{B}^6 : 45x_1 + 46x_2 + 79x_3 + 54x_4 + 53x_5 + 125x_6 \leq 178\}.$$

Then suppose we have $\hat{\mathbf{x}} \in S$, a fractional point explicitly written as

$$\hat{\mathbf{x}} = \begin{bmatrix} 0 & 0 & \frac{3}{4} & \frac{1}{2} & 1 & 0 \end{bmatrix}^T.$$

Our goal is to find a cover C which maximizes the objective $\left(\sum_{j \in C} \hat{x}_j \right) - |C| + 1$. Formulate a mixed integer linear program and solve it. Is $\hat{\mathbf{x}}$ cut off by the cover inequality you obtained from the solution?

3.24 ▪ Wheels and Tours

The wheel W_n is a graph $G = (V, E)$ with $n+1$ vertices labeled $\{0, 1, \ldots, n\}$. It has $2n$ edges of the form $\{(1,2), (2,3), \ldots, (n-1,n), (1,n)\}$ and $\{(0,i)\}$ for $i \in \{1, \ldots, n\}$. Let $c_{ij} > 0$ be the distance from node i to node j. Let d_i for $i \in \{1, \ldots, n\}$ be defined as follows:

$$d_i = \begin{cases} c_{(0,i)} + c_{(0,i+1)} - c_{(i,i+1)}, & i \in \{1, \ldots, n-1\}, \\ c_{(0,n)} + c_{(0,1)} - c_{(1,n)}, & i = n. \end{cases}$$

Explain why the following optimization problem is equivalent to an STSP (Symmetric Traveling Salesperson Problem) on G with arc lengths c_{ij}:

$$\begin{aligned} \text{minimize} \quad & \sum_{i=1}^{n} d_i y_i \\ \text{subject to} \quad & \sum_{i=1}^{n} y_i = 1, \\ & \mathbf{y} \in \mathbb{B}^n. \end{aligned} \qquad (P)$$

3.25 ▪ On Shortest Paths

Let $G = (V, E)$ be a directed graph and $\mathbf{w} \in \mathbb{Z}_+^{|E|}$. The shortest path problem can be written as follows:

$$\begin{aligned} \text{minimize} \quad & \mathbf{w}^T \mathbf{x} \\ \text{subject to} \quad & A\mathbf{x} = \begin{bmatrix} 1 & 0 & \ldots & 0 & -1 \end{bmatrix}^T, \\ & \mathbf{x} \in \mathbb{B}^{|E|}. \end{aligned} \qquad (SP)$$

In this optimization problem A is the node-arc incidence matrix, that is, a $|V| \times |E|$ dimensional matrix with entries $A_{v,(e)}$ belonging to $\{-1, 0, 1\}$.

(a) How many 0's are in A?

(b) Let $i^* \in V \setminus \{s, t\}$. Show that if $A_{i^*,(e)} \leq 0$ for any $e \in E$, then for any feasible solution $\mathbf{x}$ of (SP), $x_e = 0$ where $e \in \mathcal{E} = \{(j, i^*) | j \in \delta^-(i^*)\}$.

(c) Show that there exists no matrix $C \in \mathbb{R}^{|E| \times |V|}$ such that $AC = I_{|V|}$ where $I_{|V|}$ is the $|V|$ dimensional identity matrix.

3.26 ▪ Deriving a Cutting Plane from Covering Constraints

Let $\mathbf{x} \in \mathbb{B}^n$ satisfy the constraints

$$x_i + x_j \geq 1, \text{ for } 1 \leq i < j \leq n.$$

Show that the constraint

$$\sum_{i=1}^{n} x_i \geq n - 1$$

is valid. Give a fractional point with $\mathbf{0} \leq \mathbf{x} \leq \mathbf{1}$ that satisfies the original $\frac{n(n-1)}{2}$ constraints but violates the new constraint where $n > 2$.

3.27 ▪ Mixed Integer Rounding (MIR)

Given the mixed integer set

$$X = \{\mathbf{v} \in \mathbb{R}_+, y \in \mathbb{Z} : \mathbf{v} + y \geq 7.3\},$$

give a mixed integer rounding (MIR) inequality (see Figure 13) valid for X as well as a facet defining inequality for the convex hull of X.

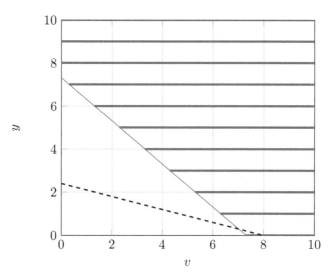

Figure 13. *An Example of an MIR Cut.*

3.28 ▪ Mixed Integer Rounding: Relax and MIR

Consider the set

$$P^1 = \left\{ \mathbf{v} \in \mathbb{R}_+^{|C|}, \mathbf{y} \in \mathbb{Z}_+^{|I|} : \sum_{j \in C} c_j v_j + \sum_{j \in I} a_j y_j \geq b \right\}.$$

Using the solution of the previous problem, obtain an MIR inequality for the set P^1.

Hint: Rewrite the inequality first by passing to a larger left-hand side that can be recognized as the general form required for an MIR inequality as in the previous problem.

3.29 ▪ Relax and MIR: Multiple Constraints

In the previous two problems you have been exposed to the technique of mixed integer rounding (MIR) for generating a valid inequality, if necessary after relaxing the constraint. The purpose of this problem is to take this procedure one step further by combining multiple constraints via non-negative multipliers (recall the Chvátal–Gomory procedure).

Consider the set

$$T = \left\{ v \in \mathbb{R}_+, x \in \mathbb{Z}_+ : \begin{array}{l} -v - 4x \geq -4, \\ -v + 4x \geq 0 \end{array} \right\}.$$

Show that $v = 0$.

Hint: After passing to equalities using surplus variables, use non-negative multipliers $\lambda_1 = -\frac{1}{8}$ and $\lambda_2 = \frac{1}{8}$ to obtain a "base" equality. Then, apply Relax and MIR.

3.30 ▪ Flow Cover Inequalities

The single-node flow set is the mixed integer linear set defined as follows:

$$T = \left\{ (\mathbf{x}, \mathbf{y}) \in \mathbb{B}^n \times \mathbb{R}_+^n : \begin{array}{l} \sum_{j=1}^{n} y_j \leq b, \\ y_j \leq a_j x_j, \quad \forall j \in \{1, \dots, n\} \end{array} \right\},$$

where $0 < a_j \leq b$ for all $j \in \{1, \dots, n\}$. This structure appears in many integer programming formulations that model fixed charges. The elements of the set T can be interpreted in terms of a network consisting of n arcs with capacities $\{a_1, \dots, a_n\}$ entering the same node, and one arc of capacity b going out. The variable x_j indicates whether arc j is open, while y_j is the flow through arc j, for $j \in \{1, \dots, n\}$.

Let $N = \{1, \dots, n\}$. A set $C \subset N$ is called a *flow cover* if $\sum_{j \in C} a_j > b$ (notice that this is the same as binary knapsack covers). Define the residual excess capacity of C as λ, where $\lambda = \sum_{j \in C} a_j - b$. Then, we can write the following inequality:

$$\sum_{j \in C} y_j + \sum_{j \in C} (a_j - \lambda)_+ (1 - x_j) \leq b$$

$((a_j - \lambda)_+ = \max(0, a_j - \lambda))$, which is a valid inequality for T. In this problem we shall apply the flow cover inequalities to the capacitated version of the facility location problem, which is formulated as

$$
\begin{array}{lll}
\text{minimize} & \displaystyle\sum_{i=1}^{m}\sum_{j=1}^{n} c_{ij} y_{ij} + \sum_{i=1}^{m} f_i x_i & \\
\text{subject to} & \displaystyle\sum_{i=1}^{m} y_{ij} = d_j, & \forall j \in \{1, \ldots, n\}, \\
& \displaystyle\sum_{j=1}^{n} y_{ij} \le u_i x_i, & \forall i \in \{1, \ldots, m\}, \\
& Y \in \mathbb{R}^m_+, & \forall j \in \{1, \ldots, n\}, \\
& \mathbf{x} \in \mathbb{B}^m.
\end{array}
\tag{P}
$$

Here, x_i is a binary variable representing whether or not a facility is open, continuous variable y_{ij} represents the amount of goods shipped from facility i to client j, d_j is the demand of client j, and u_i is the capacity of facility i.

Show how to apply flow cover inequalities to this facility location problem by generating valid flow cover inequalities.

3.31 ▪ Split Cuts

The inequality $\mathbf{c}^T\mathbf{x} + \mathbf{h}^T\mathbf{y} \le c_0$ is a *split cut* for $P \cap (\mathbb{Z}^n \times \mathbb{R}^p)$ if there exists $(\boldsymbol{\pi}, \pi_0) \in \mathbb{Z}^n \times \mathbb{Z}$ such that $\mathbf{c}^T\mathbf{x} + \mathbf{h}^T\mathbf{y} \le c_0$ is valid for the sets $P \cap \{(\mathbf{x}, \mathbf{y}) : \boldsymbol{\pi}^T\mathbf{x} \le \pi_0\}$ and $P \cap \{(\mathbf{x}, \mathbf{y}) : \boldsymbol{\pi}^T\mathbf{x} \ge \pi_0 + 1\}$.

Furthermore, if $\mathbf{c}^T\mathbf{x} + \mathbf{h}^T\mathbf{y} \le c_0$ is a split cut for $X = P \cap (\mathbb{Z}^n \times \mathbb{R}^p)$, then $\mathbf{c}^T\mathbf{x} + \mathbf{h}^T\mathbf{y} \le c_0$ is a valid inequality for X.

Consider the two variable set $X = P \cap \mathbb{Z}^2$, where $P = \{\mathbf{x} \in \mathbb{R}^2 : x_1 + 4x_2 \ge \frac{11}{4}, x_1 + x_2 \ge \frac{5}{4}\}$. Show that $4x_1 + 10x_2 \ge 11$ is a split cut for X.

Hint: It may help to draw a picture.

3.32 ▪ Packings and Integer Knapsack Cuts

Consider the following integer set:

$$
X = \left\{ \mathbf{x} \in \mathbb{Z}^4_+ : \begin{array}{l} 7x_1 + 5x_2 + 4x_3 + 3x_4 \ge 42, \\ x_i \le 3, \quad \forall i \in \{1, \ldots, 4\} \end{array} \right\}.
$$

(a) Prove that the inequality

$$
x_3 + x_4 \ge 2
$$

is a valid inequality for X. This type of inequality is referred to as an *integer knapsack cut*.

(b) Show, if necessary after solving part (a), that the following inequalities are also valid for x:

$$x_2 + x_4 \geq 2,$$
$$x_2 + x_3 \geq 3,$$
$$x_1 \geq 1.$$

3.33 ▪ Strengthening Integer Knapsack Cuts

Consider again the integer set

$$X = \left\{ \mathbf{x} \in \mathbb{Z}_+^4 : \begin{array}{l} 7x_1 + 5x_2 + 4x_3 + 3x_4 \geq 42, \\ x_i \leq 3, \quad \forall i \in \{1, \ldots, 4\} \end{array} \right\}.$$

(a) Show that the inequality $x_1 + x_3 + x_4 \geq 4$ is valid for X.

(b) Show that the previous inequality can be strengthened to

$$x_1 + x_3 + x_4 \geq 5.$$

3.34 ▪ Interval Matrices and Integral Polytopes

The following background information is useful for this problem.

An *interval matrix* M is a matrix with only $0, 1$ entries such that each row has the form

$$(0, \ldots, 0, 1, \ldots, 1, 0, \ldots, 0),$$

i.e., the ones are clustered together without interruption in each row. It may happen that an entire row of ones is also present in the matrix. An interval matrix is known to be totally unimodular (TU). It is also given that a polytope $P = \{\mathbf{x} \in \mathbb{R}^n : M\mathbf{x} \leq \mathbf{b}, \mathbf{l} \leq \mathbf{x} \leq \mathbf{u}\}$ with $\mathbf{b}, \mathbf{l}, \mathbf{u}$ integer vectors, and M a TU matrix, and all extreme points of P are integer vectors. Such polytopes are referred to as *integral polytopes*.

You are given a graph with a tree structure rooted at node 0 with T levels. Node 0 has N children at level 1; then each node at level 1 has N children, and so on. For example, for $T = 2$ and $N = 3$, we have the graph in Figure 14.

Denote by $\mathcal{N}$ the set of all nodes, e.g., $\mathcal{N} = \{0, 1, \ldots, 12\}$ in our example. The nodes $\{4, 5, \ldots, 11, 12\}$ are the *leaf* nodes in our case. We denote by $\mathcal{A}(n)$ the set of nodes in the unique *path* from the root node 0 to leaf node n. In our example, $\mathcal{A}(4) = \{0, 1, 4\}$, and $\mathcal{A}(7) = \{0, 2, 7\}$. Associate a variable x_n with each node n in

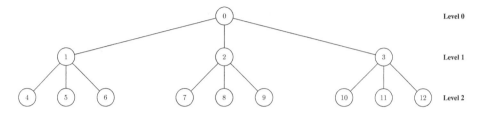

Figure 14. *A Tree Structure Rooted at Node 0 with 2 Levels.*

the tree, and consider the following problem for some positive integer k:

$$
\begin{aligned}
\text{minimize} \quad & \sum_{n \in \mathcal{N}} c_n x_n \\
\text{subject to} \quad & \sum_{m \in \mathcal{A}(n)} x_m \leq k, \quad \text{for all leaves } n, \\
& \mathbf{x} \in \mathbb{B}^{|\mathcal{N}|}.
\end{aligned}
\qquad (P)
$$

Prove that the polytope

$$
\left\{ \mathbf{x} : \begin{array}{ll} \sum_{m \in \mathcal{A}(n)} x_m \leq k, & \text{for all leaves } n, \\ x_n \in [0,1], & \forall n \in \mathcal{N} \end{array} \right\}
$$

is an integral polytope.

Hint: The transpose of a TU matrix is also TU.

3.35 ▪ Integer Programming Duality: Lagrangian Relaxation

The purpose of this problem is to introduce Lagrangian duality for integer optimization. Consider the integer optimization problem (P):

$$
\begin{aligned}
Z = \quad & \text{maximize } \mathbf{c}^T \mathbf{x} \\
& \text{subject to } A\mathbf{x} \leq \mathbf{b}, \\
& \phantom{\text{subject to }} D\mathbf{x} \leq \mathbf{e}, \\
& \phantom{\text{subject to }} \mathbf{x} \in \mathbb{Z}_+^n,
\end{aligned}
\qquad (P)
$$

where $\mathbf{b} \in \mathbb{R}^m$, $\mathbf{e} \in \mathbb{R}^k$, and all other matrices have conformable dimensions. We assume that the constraints of (P) have been partitioned into the two sets $A\mathbf{x} \leq \mathbf{b}$ and $D\mathbf{x} \leq \mathbf{e}$ so that (P) is relatively easy to solve if the constraint set $A\mathbf{x} \leq \mathbf{b}$ is removed. To create the Lagrangian problem, we first define an m dimensional vector of non-negative multipliers $\mathbf{u}$ and add the non-negative term $\mathbf{u}^T(\mathbf{b} - A\mathbf{x})$ to the objective function of (P) to obtain

$$
\begin{aligned}
\text{maximize} \quad & \mathbf{c}^T \mathbf{x} + \mathbf{u}^T(\mathbf{b} - A\mathbf{x}) \\
\text{subject to} \quad & A\mathbf{x} \leq \mathbf{b}, \\
& D\mathbf{x} \leq \mathbf{e}, \\
& \mathbf{x} \in \mathbb{Z}_+^n.
\end{aligned}
$$

It is clear that the optimal value of this problem for u fixed at non-negative values is an upper bound on Z because we have merely added a non-negative term to the objective function. At this point, we create the Lagrangian problem LR_u by removing the constraints $A\mathbf{x} \le \mathbf{b}$ to get the Lagrangian function:

$$Z_D(\mathbf{u}) = \quad \text{maximize } \mathbf{c}^T\mathbf{x} + \mathbf{u}^T(\mathbf{b} - A\mathbf{x})$$
$$\text{subject to } D\mathbf{x} \le \mathbf{e},$$
$$\mathbf{x} \in \mathbb{Z}_+^n.$$

Since removing the constraints $A\mathbf{x} \le \mathbf{b}$ cannot decrease the optimal value, $Z_D(\mathbf{u})$ is also an upper bound on Z. Moreover, by assumption computing $Z_D(\mathbf{u})$ (i.e., solving the Lagrangian subproblem) should be easier compared to computing Z. The integer programming dual is defined to be the problem

$$\text{minimize} \quad Z_D(\mathbf{u})$$
$$\text{subject to} \quad \mathbf{u} \in \mathbb{R}_+^m.$$

Now consider the following example problem $(P1)$:

$$\text{maximize} \quad 16x_1 + 10x_2 + 4x_4$$
$$\text{subject to} \quad 8x_1 + 2x_2 + x_3 + 4x_4 \le 10,$$
$$x_1 + x_2 \le 1, \qquad\qquad (P1)$$
$$x_3 + x_4 \le 1,$$
$$\mathbf{x} \in \mathbb{Z}_+^4.$$

If we dualize constraint "$8x_1 + 2x_2 + x_3 + 4x_4 \le 10$," we get the Lagrangian problem

$$Z_D(u) = \quad \text{maximize } 16x_1 + 10x_2 + 4x_4 + u(10 - 8x_1 - 2x_2 - x_3 - 4x_4)$$
$$\text{subject to } x_1 + x_2 \le 1,$$
$$x_3 + x_4 \le 1,$$
$$\mathbf{x} \in \mathbb{Z}_+^4.$$

We can rewrite the above problem as

$$Z_D(u) = 10u + \quad \text{maximize } (16 - 8u)x_1 + (10 - 2u)x_2 - ux_3 + (4 - 4u)x_4$$
$$\text{subject to } x_1 + x_2 \le 1,$$
$$x_3 + x_4 \le 1,$$
$$\mathbf{x} \in \mathbb{Z}_+^4.$$

The last problem is quite easy to solve for a fixed non-negative value of u equal to one, x_1, or x_2, whichever one has the highest objective function coefficient, and do the same for x_3, x_4. Solve the dual problem $\min_{u \in \mathbb{R}_+} Z_D(u)$ by trying out different values of u.

3.36 ▪ An Improved Lagrangian Dual

This problem is a continuation of the previous problem. Consider the following Lagrangian dual instead of the one in the previous problem (with multipliers $v_1 \geq 0$, $v_2 \geq 0$):

$$Z_D(v_1, v_2) = \quad \text{maximize } 16x_1 + 10x_2 + 4x_4 + v_1(1 - x_1 - x_2) + v_2(1 - x_3 - x_4)$$
$$\text{subject to } 8x_1 + 2x_2 + x_3 + 4x_4 \leq 10,$$
$$\mathbf{x} \in \mathbb{Z}_+^4,$$

or, equivalently

$$Z_D(v_1, v_2) = v_1 + v_2 + \quad \text{maximize } (16 - v_1)x_1 + (10 - v_1)x_2 - v_2x_3 + (4 - v_2)x_4$$
$$\text{subject to } 8x_1 + 2x_2 + x_3 + 4x_4 \leq 10,$$
$$\mathbf{x} \in \mathbb{Z}_+^4.$$

(a) As in the previous problem, solve by trial and error the Lagrangian dual problem

$$\text{minimize} \quad Z_D(v_1, v_2)$$
$$\text{subject to} \quad v_1, v_2 \in \mathbb{R}_+.$$

(b) What do you observe with respect to the previous problem?

The Subgradient Method

In fact, we could arrive at an optimal solution of the Lagrangian dual not by trial and error, but using the so-called *subgradient method*. The subgradient method is a straightforward adaptation of the gradient method in which gradients are replaced by subgradients. Given an initial value $\mathbf{u}_0$ a sequence $\{\mathbf{u}_k\}$ is generated by the rule

$$\mathbf{u}_{k+1} = \max\{\mathbf{0}, \mathbf{u}_k - t_k(\mathbf{b} - A\mathbf{x}_k)\}$$

where $\mathbf{x}_k$ is an optimal solution to $(LR_{\mathbf{u}_k})$ and t_k is a positive scalar step size. A formula for choosing t_k that has proven effective in practice is

$$t_k = \frac{\lambda_k(Z_D(\mathbf{u}_k) - Z^*)}{\|\mathbf{b} - A\mathbf{x}_k\|_2^2}$$

where λ_k is a scalar satisfying $0 < \lambda_k \leq 2$ and Z^* is the value of the best known feasible solution to (P), frequently obtained by applying a heuristic to (P). Often the sequence λ_k is determined by setting $\lambda_0 = 2$ and halving λ_k whenever $Z_D(\mathbf{u})$ has failed to decrease in some fixed number of iterations.

Compute the solution of the dual problem $Z_D(u)$ in Problem 3.35 starting the subgradient iteration with $u = 0$, and $t = 1$ and

(c) halving the t value at each step,

(d) using the rule for setting λ_k given above (use 16 as the value of the best known feasible solution, i.e., use $Z^* = 16$ in the formula).

Compare the results with the bounds obtained in sections 3.35 and 3.36.

3.37 ▪ Simple Plant Location and Lagrangian Dual

Consider applying the Lagrangian dual method to the Uncapacitated Facility Location (a.k.a., Simple Plant Location) problem. Recall that we have two sets of decision variables:

$$y_j = \begin{cases} 1, & \text{warehouse at site } j \text{ is open,} \\ 0 & \text{otherwise.} \end{cases}$$

$$x_{ij} = \begin{cases} 1, & \text{warehouse at site } j \text{ services customer } i, \\ 0 & \text{otherwise.} \end{cases}$$

Then we have the objective

$$Z_{UFL} = \min \sum_{j=1}^{m} f_j y_j + \sum_{i=1}^{n} \sum_{j=1}^{m} c_{ij} x_{ij}$$

with the following constraints.

• Every customer must be served:

$$\sum_{j=1}^{m} x_{ij} = 1, \quad \forall i \in \{1, \ldots, n\}.$$

• We cannot serve any customer from site j if no warehouse was built at site j:

$$x_{ij} \leq y_j, \quad \forall i \in \{1, \ldots, n\}, \quad j \in \{1, \ldots, m\}.$$

Consider the following Lagrangian subproblem obtained by dualizing the first set of constraints:

$$Z_D(\lambda) = \quad \text{minimize} \quad \sum_{j=1}^{m} f_j y_j + \sum_{i=1}^{n} \sum_{j=1}^{m} c_{ij} x_{ij} + \sum_{i=1}^{n} \lambda_i (1 - \sum_{j=1}^{m} x_{ij})$$

$$\text{subject to} \quad x_{ij} \leq y_j, \quad \forall i \in \{1, \ldots, n\}, \quad \forall j \in \{1, \ldots, m\},$$

$$X \in \mathbb{B}^{n \times m},$$

$$\mathbf{y} \in \mathbb{B}^{m},$$

where $\lambda_i \in \mathbb{R}$, for $i \in \{1, \dots, n\}$ (for equality constraints the dual multipliers are unrestricted in sign).

$$\text{Show that } Z_D(\boldsymbol{\lambda}) = \sum_{i=1}^{n} \lambda_i + \sum_{j=1}^{m} \min \left\{ 0, \sum_{i=1}^{n} \min[c_{ij} - \lambda_i, 0] + f_j \right\}.$$

3.38 ▪ A Dice Puzzle Solved by Dynamic Programming

Suppose you have n identical k-sided dice. Formulate a dynamic programming problem to find out how many ways to obtain m in total, where m is a positive integer. Solve it for $n = 3$, $k = 6$, and $m = 10$.

3.39 ▪ Minimum Cut versus Maximum Cut

We know that the Maximum Flow Problem can be solved as a linear optimization problem (when the capacities are integers). We also know that there are weak and strong duality relations between the Maximum Flow and Minimum Cut Problems. So if the Maximum Flow Problem is an LP, then the minimum cut problem must also be an LP! What kind of LP is it? It must be the LP dual of the Maximum Flow Problem!

Let s and t be source and sink nodes, respectively. Associating dual variables w_s, w_t, and w_j, $j \in \{1, \dots, n\}$, to flow balance equations, and non-negative variables y_{ij} to the constraints $x_{ij} \leq u_{ij}$, we obtain

$$\begin{array}{ll}
\text{minimize} & \sum_{(i,j) \in A} u_{ij} y_{ij} \\
\text{subject to} & y_{ij} \geq w_j - w_i, \quad \forall (i,j) \in A, \\
& w_t - w_s = 1, \\
& y_{ij} \in \mathbb{R}_+, \qquad \forall (i,j) \in A.
\end{array}$$

We have the following fact that is easy to see: Given a cut S, T in G, we can associate with it a feasible solution to the dual LP as follows:

$$\begin{array}{ll}
w_i & = 0, \quad \forall i \in S, \\
w_j & = 1, \quad \forall j \in T, \\
y_{ij} & = 0, \quad \forall (i,j) \in A \text{ such that } i \in S, j \in S, \text{or } i \in T, j \in T, \\
y_{ij} & = 1, \quad \forall (i,j) \in A \text{ such that } i \in S, j \in T.
\end{array}$$

Therefore, any cut S, T in the graph G can be mapped to a feasible solution of the LP above. Notice that the objective value of the solution is equal to the capacity of the cut. The converse of this statement is also true, although showing this is not the purpose of the problem. Given any feasible solution to the LP dual with value v, one can find a cut S, T and an associated cut solution $w(S), y(S)$ with value no more than v.

The aim is to pass to a maximum capacity (or maximum cardinality) cut problem formulation. Now, let us look into that problem, which is very close to the Minimum Cut Problem: given an undirected graph $G = (V, E)$, partition the set of nodes V into two subsets (non-overlapping) V_1 and V_2 ($V_1 \cup V_2 = V$) such that the number of edges between V_1 and V_2 (i.e., the number of edges in the cut) is a maximum. This problem is known as the *Maximum Cut Problem*. If the edges of the graph have (positive) weights c_{ij}, then we wish to find the cut with the maximum weight. Give an integer programming formulation for the Maximum Cut Problem.

3.40 ▪ The Cutting Stock Problem and Column Generation

A paper mill produces large rolls of paper of width W, which are then cut into rolls of various smaller widths in order to meet demand. Let m be the number of different widths that the mill produces. The mill receives an order for b_i rolls of width w_i for $i \in \{1, \ldots, m\}$, where $w_i \leq W$. How many of the large rolls are needed to meet the order? How can we formulate this problem using integer variables? Assume that an upper bound p is known on the number of large rolls to be used. Define variables y_j for $j \in \{1, \ldots, n\}$, which take value 1 if large roll j is used and 0 otherwise and variables z_{ij}, for $i \in \{1, \ldots, m\}, j \in \{1, \ldots, p\}$, indicate the number of small rolls of width w_i to be cut out of large roll j.

The formulation is

$$
\begin{aligned}
\text{minimize} \quad & \sum_{j=1}^{p} y_j \\
\text{subject to} \quad & \sum_{i=1}^{m} w_i z_{ij} \leq W y_j, \quad \forall j \in \{1, \ldots, p\}, \\
& \sum_{j=1}^{p} z_{ij} \geq b_i, \qquad \forall i \in \{1, \ldots, m\}, \\
& Z \in \mathbb{Z}_{+}^{m \times n}, \\
& \mathbf{y} \in \mathbb{B}^{p}.
\end{aligned}
\qquad (I)
$$

The objective function is to minimize the number of large rolls to be used. The first constraint expresses the requirement that the total width cut from large roll j cannot exceed the total width W of the large roll. The second constraint says that the number of small rolls of width w_i cut from large rolls should meet the demand b_i. Computational experience shows that this is not a strong formulation. The bound provided by the LP relaxation is rather distant from the optimal integer value. A better formulation is needed.

Consider all the possible different cutting patterns. Each pattern is represented by a vector $\mathbf{s} \in \mathbb{Z}^m$ where component i represents the number of rolls of width w_i cut out of the large roll. The set of cutting patterns is therefore

$$S := \left\{ \mathbf{s} \in \mathbb{Z}^m : \sum_{i=1}^m w_i s_i \leq W, \mathbf{s} \geq 0 \right\}.$$

Notice that this is an integer knapsack constraint! For example, let $W = 5$, and the order has rolls of 3 different widths $w_1 = 2.1, w_2 = 1.8$, and $w_3 = 1.5$. A possible cutting pattern consists of 3 rolls of width 1.5, i.e., we have $(0, 0, 3)$. Another consists of one roll of width 2.1 and one of width 1.8, i.e., we have $(1, 1, 0)$. Yet another one is to have 2 rolls of width 1.5 and one roll of width 1.8 resulting in the pattern $(0, 1, 2)$. Obviously it would be a lot of work to form the entire set S of cutting patterns. Introduce integer variables $x_{\mathbf{s}}$ representing the number of rolls cut according to pattern $\mathbf{s} \in S$. The second formulation is

$$
\begin{aligned}
\text{minimize} \quad & \sum_{\mathbf{s} \in S} x_{\mathbf{s}} \\
\text{subject to} \quad & \sum_{\mathbf{s} \in S} s_i x_{\mathbf{s}} \geq b_i, \quad i \in \{1, \ldots, m\}, \qquad (II) \\
& \mathbf{x} \in \mathbb{Z}_+^{|S|}.
\end{aligned}
$$

This is an integer programming formulation in which the columns of the constraint matrix are all the feasible solutions of a knapsack set. The number of these columns (i.e., the number of possible patterns) is typically enormous, but this is a strong formulation as the LP relaxation bound is usually close to the optimum integer value. There is a potential problem though: one cannot possibly find all the patterns beforehand and formulate the problem. The solution is to start with a subset of the patterns and generate new patterns from the set S as needed using LP duality. The resulting scheme is called *column generation*. Consider the LP relaxation of formulation (II):

$$
\begin{aligned}
\text{minimize} \quad & \sum_{\mathbf{s} \in S} x_{\mathbf{s}} \\
\text{subject to} \quad & \sum_{\mathbf{s} \in S} s_i x_{\mathbf{s}} \geq b_i, \quad i \in \{1, \ldots, m\}, \\
& \mathbf{x} \in \mathbb{R}_+^{|S|}.
\end{aligned}
$$

The dual of the LP relaxation is:

$$
\begin{aligned}
\text{maximize} \quad & \sum_{i=1}^m b_i u_i \\
\text{subject to} \quad & \sum_{i=1}^m s_i u_i \leq 1, \quad \forall \mathbf{s} \in S, \\
& \mathbf{u} \in \mathbb{R}_+^m.
\end{aligned}
$$

Let S' be a subset of S, and consider the cutting stock problem formulation (II) restricted to the variables indexed by S'. The dual is the problem defined by the inequalities above indexed by S'. Let $\tilde{\mathbf{x}}, \tilde{\mathbf{u}}$ be optimal solutions to the LP relaxations above restricted to S'. By setting $\tilde{x}_\mathbf{s} = 0, \mathbf{s} \in S \setminus S'$, $\tilde{\mathbf{x}}$ can be extended to a feasible solution of the linear relaxation. By strong duality $\tilde{\mathbf{x}}$ is an optimal solution of the linear relaxation if $\tilde{\mathbf{u}}$ provides a feasible solution to the dual (defined over S). The solution $\tilde{\mathbf{u}}$ is feasible for the dual if and only if $\sum_{i=1}^{m} s_i \tilde{u}_i \leq 1$ for every $\mathbf{s} \in S$. Equivalently, the solution is feasible for the dual if and only if the value of the knapsack problem

$$\max \left\{ \sum_{i=1}^{m} s_i \tilde{u}_i : \mathbf{s} \in S \right\}$$

is equal to at most 1. If the value of this knapsack problem exceeds 1, defining $\mathbf{s}^*$ to be an optimal solution, then $\mathbf{s}^*$ corresponds to a constraint of the dual LP that is most violated by $\tilde{\mathbf{u}}$, and $\mathbf{s}^*$ is added to S', thus enlarging the set of candidate patterns. This is the column generation scheme, where variables of a linear program with many variables (exponentially many) are generated as needed.

Suppose this mill manufactures rolls of paper of a standard width of 3 meters. But customers want to buy paper rolls of shorter width, and the mill has to cut such rolls from the 3 m rolls. One 3 m roll can be cut, for instance, into two rolls 93 cm wide, one roll of width 108 cm, and the rest of 6 cm (which goes to waste). Consider an order of

- 97 rolls of width 135 cm,

- 610 rolls of width 108 cm,

- 395 rolls of width 93 cm, and

- 211 rolls of width 42 cm.

(a) What is the smallest number of 3 m rolls that have to be cut in order to satisfy this order, and how should they be cut?

(b) Starting with a subset S' of patterns with 3 elements, use the column generation scheme to optimally solve the problem above.

Part II

Solutions

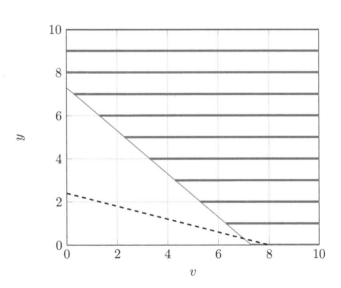

Chapter 1

Mathematical Modeling Problems with Integer Variables

1.1 ▪ Greatest Common Divisor

The model is a direct result of the definition:

$$
\begin{aligned}
\text{minimize} \quad & ax + by \\
\text{subject to} \quad & ax + by \geq 1, \\
& x \in \mathbb{Z}, \\
& y \in \mathbb{Z}.
\end{aligned}
$$

1.2 ▪ Bin Packing

We have the following model:

$$
\begin{aligned}
\text{minimize} \quad & z \\
\text{subject to} \quad & \sum_{j=1}^{n} x_{ij} = 1, && \forall i \in \{1, \ldots, n\}, \text{ (every item to a bin)} \\
& \sum_{i=1}^{n} a_i x_{ij} \leq b y_j, && \forall j \in \{1, \ldots, n\}, \text{ (capacity constraint for a used bin)} \\
& z \geq b y_j - \sum_{i=1}^{n} a_i x_{ij}, && \forall j \in \{1, \ldots, n\}. \text{ (unused total space)} \\
& X \in \mathbb{B}^{n \times n}, \\
& \mathbf{y} \in \mathbb{B}^{n}.
\end{aligned}
$$

1.3 ▪ Three Cups Puzzle

Let $i \in \{1, \ldots, 6\}$ and $j \in \{A, B, C\}$. We introduce the following set of variables:

$$y_i = \begin{cases} 1 & \text{if the player made an action at move } i, \\ 0 & \text{otherwise.} \end{cases}$$

$$x_{ij} = \begin{cases} 1 & \text{if the player made an action on cup } j \text{ at move } i, \\ 0 & \text{otherwise.} \end{cases}$$

According to these we can write the following model:

$$\begin{aligned}
\text{minimize} \quad & \sum_{i=1}^{6} y_i \\
\text{subject to} \quad & \sum_{i=1}^{6} x_{iA} = 1 + 2k_A, \\
& \sum_{i=1}^{6} x_{iB} = 2k_B, \\
& \sum_{i=1}^{6} x_{iC} = 1 + 2k_C, \\
& \sum_{j \in \{A,B,C\}} x_{ij} = 2y_i, \quad \forall i \in \{1, \ldots, 6\}, \\
& X \in \mathbb{B}^{6 \times 3}, \\
& \mathbf{y} \in \mathbb{B}^6, \\
& \mathbf{k} \in \mathbb{Z}_+^3.
\end{aligned}$$

1.4 ▪ Table Tennis Tournament

Define the following variables:

$$x(i, j, k, l) = \begin{cases} 1, & \text{player } k \text{ from first team plays with player } l \text{ at day } i \text{ and time block } j, \\ 0 & \text{otherwise.} \end{cases}$$

$$\begin{aligned}
& \sum_{i=1}^{4} \sum_{j=1}^{4} x(i, j, k, l) = 1, \quad \forall k, l \in \{1, \ldots, 4\}, \\
& \sum_{k=1}^{4} \sum_{l=1}^{4} x(i, j, k, l) = 1, \quad \forall i, j \in \{1, \ldots, 4\}, \\
& \sum_{i=1}^{4} \sum_{k=1}^{4} x(i, j, k, l) = 1, \quad \forall j, l \in \{1, \ldots, 4\}, \\
& \sum_{i=1}^{4} \sum_{l=1}^{4} x(i, j, k, l) = 1, \quad \forall j, k \in \{1, \ldots, 4\}, \\
& \sum_{j=1}^{4} \sum_{k=1}^{4} x(i, j, k, l) = 1, \quad \forall i, l \in \{1, \ldots, 4\}, \\
& \sum_{j=1}^{4} \sum_{l=1}^{4} x(i, j, k, l) = 1 \quad \forall i, k \in \{1, \ldots, 4\}.
\end{aligned}$$

The first equality is for the first bullet, the second is for the second bullet, and the remaining four are for the last bullet. An arbitrary choice of the objective will work for the model since it is a feasibility problem.

This problem is about a Round-robin tournament. Two related sources are https://www.econstor.eu/bitstream/10419/147668/1/manuskript_613.pdf and https://link.springer.com/content/pdf/10.1007%2Fb11828.pdf

1.5 ▪ Floor and Ceiling Operators

$\lfloor x \rfloor = k_x$ iff $x - 1 < k_x \leq x$ and $\lceil y \rceil = k_y$ iff $y < k_y \leq y + 1$. Then for a sufficiently small ϵ we may write

$$
\begin{aligned}
\text{maximize} \quad & x + y \\
\text{subject to} \quad & k_x + k_y = 5, \\
& x - 1 + \epsilon \leq k_x \leq x, \\
& y + \epsilon \leq k_y \leq y + 1, \\
& x, y \geq -5, \\
& k_x, k_y \in \mathbb{Z}.
\end{aligned}
\qquad \text{(MIP)}
$$

1.6 ▪ A Chemical Reaction

This problem is adapted from work by Sen, Agarwal, and Sen [18]. We write the following model:

$$
\begin{aligned}
\text{minimize} \quad & x_1 + x_2 + x_3 + x_4 + x_5 \\
\text{subject to} \quad & x_1 = 2x_3, \\
& x_1 = 2x_5, \\
& 3x_1 = 3x_3 + x_4, \\
& x_2 = x_3 + x_4, \\
& x_i \geq 1, \qquad\qquad \forall i \in \{1, \ldots, 5\}, \\
& \mathbf{x} \in \mathbb{Z}^5.
\end{aligned}
$$

Each equality constraint conserves the number of K, N, O, C atoms, respectively. If we let

$$
A = \begin{bmatrix}
1 & 0 & -2 & 0 & 0 \\
1 & 0 & 0 & 0 & -2 \\
3 & 0 & -3 & -1 & 0 \\
0 & 1 & -1 & -1 & 0
\end{bmatrix}
$$

then our problem turns out to be

$$\begin{aligned} \text{minimize} \quad & \mathbf{1}^T \mathbf{x} \\ \text{subject to} \quad & A\mathbf{x} = \mathbf{0}, \\ & x_i \geq 1, \ \forall i \in \{1, \dots 5\}, \\ & \mathbf{x} \in \mathbb{Z}^5. \end{aligned}$$

After reduced row echelon form, one can see that null space of A contains vectors of the form $\begin{bmatrix} 2x_5 & 4x_5 & x_5 & 3x_5 & x_5 \end{bmatrix}^T$.

Thus one can write an equivalent optimization problem $\min 11x_5$ st. $x_5 \geq 1$, and it has a unique optimal solution, $x_5 = 1$. Since there is 1-1 correspondence between feasible sets of these equivalent problems, optimal solution of the original problem is unique too.

1.7 ▪ Largest k Elements of a Vector

(a) This is a special case of Binary Knapsack.

$$\begin{aligned} \text{maximize} \quad & \sum_{i=1}^{n} v_i x_i \\ \text{subject to} \quad & \sum_{i=1}^{n} x_i \leq k, \\ & \mathbf{x} \in \mathbb{B}^n. \end{aligned}$$

(b) First we write the non-linear version of the problem using the positive part function, i.e., $x^+ = \max\{x, 0\}$,

$$\begin{aligned} \text{minimize} \quad & k \max_{i=1,\dots,n} \{(v_i - y_i)^+\} + \sum_{i=1}^{n} y_i \\ \text{subject to} \quad & \mathbf{y} \in \mathbb{Z}_+^n. \end{aligned}$$

Its linearization is the following:

$$\begin{aligned} \text{minimize} \quad & ky_0 + \sum_{i=1}^{n} y_i \\ \text{subject to} \quad & y_0 + y_i \geq v_i, \quad \forall i \in \{1, \dots, n\}, \\ & \mathbf{y} \in \mathbb{Z}_+^n, \\ & y_0 \in \mathbb{Z}_+. \end{aligned}$$

(c) Linear relaxations can be written in matrix form.

$$\begin{aligned} \text{maximize} \quad & \mathbf{v}^T \mathbf{x} \\ \text{subject to} \quad & \begin{bmatrix} \mathbf{1}^T \\ \mathbb{I} \end{bmatrix} x \leq \begin{bmatrix} k \\ \mathbf{1} \end{bmatrix}, \\ & \mathbf{x} \in \mathbb{R}_+^n. \end{aligned}$$

$$\text{minimize} \quad \begin{bmatrix} k & \mathbf{1}^T \end{bmatrix} \begin{bmatrix} y_0 \\ \mathbf{y} \end{bmatrix}$$

$$\text{subject to} \quad \begin{bmatrix} \mathbf{1} & \mathbb{I} \end{bmatrix} \begin{bmatrix} y_0 \\ \mathbf{y} \end{bmatrix} \geq \mathbf{v},$$

$$\mathbf{y} \in \mathbb{R}_+^n,$$

$$y_0 \in \mathbb{R}_+.$$

Using the hint we conclude that they have a primal-dual relation.

1.8 ▪ Coloring Nodes of a Graph and Classroom Assignment

We map the problem into a graph node coloring instance. Let K denote the set of colors, which is equivalent to the set of distinct classrooms, and let L denote the set of all sections, which is equivalent to set of nodes in the graph. We let two sections p and q be joined by a link (p, q) if they share an instructor or have been scheduled at the same time slot.

Let us define

$$x_{ij} = \begin{cases} 1 & \text{if color } j \text{ is used for section } i, \\ 0 & \text{otherwise.} \end{cases}$$

$$y_j = \begin{cases} 1 & \text{if color } j \text{ is used,} \\ 0 & \text{otherwise.} \end{cases}$$

Then we minimize the number of colors used:

$$\min \sum_{j \in K} y_j$$

subject to

$$\sum_{j \in K} x_{ij} = 1, \quad \forall i \in L,$$

(every section has to be assigned a classroom)

$$x_{ij} + x_{kj} \leq y_j, \quad \forall j \in K, \quad k \in N(i) \cup T(i), \quad i \in L$$

(for every section i, the sections having a common schedule or instructor cannot be in the same classroom, or two neighbor nodes cannot receive the same color).

1.9 ▪ A Chessboard Problem

Let $i, j \in \{1, \ldots, 8\}$. We introduce the following set of variables:

$$v_{i,j} = \begin{cases} 1 & \text{if there exists a domino tile occupying } (i,j) \text{ and } (i+1,j), \\ 0 & \text{otherwise.} \end{cases}$$

$$h_{i,j} = \begin{cases} 1 & \text{if there exists a domino tile occupying } (i,j) \text{ and } (i,j+1), \\ 0 & \text{otherwise.} \end{cases}$$

Then we have the following model:

$$\text{maximize} \quad \sum_{i=1}^{8} \sum_{j=1}^{8} (h_{i,j} + v_{i,j})$$

$$\text{subject to} \quad \sum_{i=1}^{8} h_{i,8} + \sum_{j=1}^{8} v_{8,j} = 0,$$

$$6(1 - v_{i,j}) \geq v_{i+1,j} + v_{i-1,j} + h_{i,j} + h_{i,j-1} + h_{i+1,j} + h_{i+1,j-1}, \forall i, j \in \{1, \ldots, 8\},$$

$$6(1 - h_{i,j}) \geq h_{i,j+1} + h_{i,j-1} + v_{i,j} + v_{i-1,j} + v_{i,j+1} + v_{i-1,j+1}, \forall i, j \in \{1, \ldots, 8\},$$

$$H, V \in \mathbb{B}^{8 \times 8}.$$

Indices exceeding bounds can be arranged by enlarging the board and setting all $v(0, j) = v(9, j) = h(i, 0) = h(i, 9) = 0$ for all i, j.

An initial upper bound on the optimal value can be found by $\frac{64}{2} = 32$; indeed it is attainable by a simple arrangement. For the mutilated board, an initial upper bound is $\frac{62}{2} = 31$. Each domino tile occupies a white and black tile. WLOG, removing tiles from opposite corners results in 30 white and 32 black tiles. Therefore the upper bound for this problem is 30, and it is attainable by a simple arrangement.

1.10 ▪ The 8-Queens

Let i, j be indices from 1 to 8; then

$$x_{ij} = \begin{cases} 1 & \text{if there is a queen placed on cell } (i,j), \\ 0 & \text{otherwise.} \end{cases}$$

One should maximize the number of queens; hence the objective will be

$$\max \sum_{i=1}^{8} \sum_{j=1}^{8} x_{ij}.$$

If there exists a queen on a cell, there must not be any queens in the same row, column, or diagonal; hence

$$64(1 - x_{ij}) \geq \sum_{k=1}^{8} \sum_{l=1}^{8} C_{ij}(k,l)x_{kl}, \quad \forall i,j \in \{1,\ldots,8\},$$

where C_{ij} is a parameter defined as follows:

$$C_{ij}(k,l) = \begin{cases} 1 & \text{if } (i,j) \text{ and } (k,l) \text{ are on the same row, column, or diagonal} \\ & \text{and } (i,j) \neq (k,l), \\ 0 & \text{otherwise.} \end{cases}$$

1.11 ▪ A Basket of Apples

Let x_i be 1 if the ith apple is in the first basket and 0 otherwise. Then one can write the following model:

$$\begin{aligned} \text{minimize} \quad & t \\ \text{subject to} \quad & -t \leq \sum_{i=1}^{m} w_i x_i - \sum_{i=1}^{m} w_i(1 - x_i) \leq t, \\ & \sum_{i=1}^{m} w_i x_i \leq C, \\ & \sum_{i=1}^{m} w_i(1 - x_i) \leq C, \\ & \mathbf{x} \in \mathbb{B}^m, \\ & t \in \mathbb{R}. \end{aligned}$$

For the general case we need to define

$$x_{ij} = \begin{cases} 1 & \text{if the } i\text{th apple is put into } j\text{th basket,} \\ 0 & \text{otherwise.} \end{cases}$$

Then one can write the following model:

$$\begin{aligned} \text{minimize} \quad & \sum_{j=1}^{n} \sum_{k=1}^{n} t_{jk} \\ \text{subject to} \quad & -t_{jk} \leq \sum_{i=1}^{m} w_i x_{ij} - \sum_{i=1}^{m} w_i x_{ik}, \quad \forall j,k \in \{1,\ldots,n\}, \\ & \sum_{i=1}^{m} w_i x_{ij} - \sum_{i=1}^{m} w_i x_{ik} \leq t_{jk}, \quad \forall j,k \in \{1,\ldots,n\}, \\ & \sum_{i=1}^{m} w_i x_{ij} \leq C_j, \quad \forall j \in \{1,\ldots,n\}, \\ & \sum_{j=1}^{n} x_{ij} = 1, \quad \forall i \in \{1,\ldots,m\}, \\ & X \in \mathbb{B}^{m \times n}, \\ & T \in \mathbb{R}_+^{n \times n}. \end{aligned}$$

1.12 ▪ A Grocery Chain

We define x_i to check whether a store is opened at location i and y_{ij} is the number of units served from location i to customer j.

$$
\begin{aligned}
\text{minimize} \quad & \sum_{i=1}^{n} B_i x_i + \sum_{i=1}^{n} \sum_{j=1}^{m} c_{ij} y_{ij} + \sum_{j=1}^{m} p_j t_j^- \\
\text{subject to} \quad & y_{ij} \le M x_i, && \forall i \in \{1, \ldots, n\}, \ \forall j \in \{1, \ldots, m\}, \\
& d_j - \sum_{i=1}^{n} y_{ij} = t_j, && \forall j \in \{1, \ldots, m\}, \\
& t_j = t_j^+ - t_j^-, && \forall j \in \{1, \ldots, m\}, \\
& \sum_{j=1}^{m} y_{ij} \le s_i, && \forall i \in \{1, \ldots, n\}, \\
& Y \in \mathbb{R}_+^{n \times m}, \\
& \mathbf{x} \in \mathbb{B}^n, \\
& t, t^+, t^- \in \mathbb{R}_+.
\end{aligned}
$$

In addition to these we add

$$
\begin{aligned}
& x_1 + x_4 \ge 1, \\
& \sum_{i=1}^{n} x_i \le 6, \\
& x_1 + x_3 + x_5 \le 2 + w, \\
& 2(1 - w) \le x_7 + x_8, \\
& w \in \mathbb{B}.
\end{aligned}
$$

1.13 ▪ An Infeasible Integer Program

$$
\begin{aligned}
\text{minimize} \quad & \mathbf{c}^T \mathbf{x} + \sum_{i=1}^{m} d y_i \\
\text{subject to} \quad & \mathbf{a}_i^T \mathbf{x} \le b_i + M_i y_i, && \forall i \in \{1, \ldots, m\}, \\
& x_j \le 1, && \forall j \in \{1, \ldots, n\}, \\
& \mathbf{y} \in \mathbb{B}^m, \\
& \mathbf{x} \in \mathbb{R}_+^n.
\end{aligned}
$$

For this formulation we use $M_i = \max_{0 \le \mathbf{x} \le 1} \{\mathbf{a}_i^T \mathbf{x} - b_i\}$ for all $i \in \{1, \ldots, m\}$.

1.14 ▪ Packaging Chicken Nuggets

$$
\begin{aligned}
n &= \text{number of different package types}, \\
s_j &= \text{size of } j\text{th type of package for } j \in \{1, \ldots, n\}, \\
m &= \min_{j \in \{1, \ldots, n\}} s_j.
\end{aligned}
$$

According to these parameters we define variables

$$k = \text{number of pieces ordered,}$$

$$x_{ij} = \text{number of packages needed from type } j \text{ to meet } k + i \text{ pieces of order.}$$

Then we have

$$
\begin{aligned}
\text{minimize} \quad & k \\
\text{subject to} \quad & \sum_{j=1}^{n} s_j x_{ij} = k + i, \quad \forall i \in \{1, \ldots, m\}, \\
& X \in \mathbb{Z}_+^{m \times n}, \\
& k \in \mathbb{Z}_+.
\end{aligned}
$$

1.15 ▪ Handling Unions of Polyhedra

$$
\begin{aligned}
\text{minimize} \quad & 3x + 2y \\
\text{subject to} \quad & x \leq 4 + z, \\
& x \geq 2 + z, \\
& y \leq 1 + 3z, \\
& y \geq 3z, \\
& x, y \in \mathbb{R}, \\
& z \in \mathbb{B}.
\end{aligned}
$$

1.16 ▪ Simplifying an Integer Program

Let $x = x_1 + x_2$ and $y = x_3$. Then the problem can be simplified up to

$$
\begin{aligned}
\text{maximize} \quad & ax + by \\
\text{subject to} \quad & x + y \leq 1, \\
& x \in \mathbb{Z}_+, \\
& y \in \mathbb{Z}.
\end{aligned}
$$

If we rewrite the objective as $a(x + y) + (b - a)y$, this expression is upper bounded by $a + (b - a)y$. We also know that $y \leq 1$. Hence if $b \geq a$, the objective is bounded by b; otherwise it is unbounded since y can be taken as large as possible (in the sense of magnitude) to make the objective $+\infty$.

1.17 ▪ Equivalence of Integer Sets

The smallest set $G = \{(-1, 1), (0, 1), (1, 1), (-1, 0), (0, 0), (1, 0), (0, -1)\}$. G can be found by searching around the origin, using the hint. Therefore we can define

$P = \operatorname{conv}(G) = \{(x, y) \in \mathbb{R}^2 : y \le 1, -1 \le x \le 1, x - y \le 1, -x - y \le 1\}$.
Optimizing over $H \cap \mathbb{Z}^2$ is equivalent to optimizing on G; hence $(1, 0)$ is an optimal solution with optimal value 3 .

Assume $x^2 + y^2 = 2 + \epsilon$ for some $\epsilon > 0$. Then we have to show that

$$(x^2 + y^2 - 1)^3 > x^2 y^3.$$

Assume we have the following polynomial;

$$p(y) = y^6 - (2 + \epsilon)y^3 + (\epsilon + 1)^3.$$

For $y = 0$, we have $p(y) > 0$. Since $g(y) = y^3$ is a bijection with $g(y)$ and y having equal signs, we can say that p is a quadratic polynomial in terms of g. Then we check $\triangle p = -8\epsilon - 8\epsilon^2 - 4\epsilon^3 < 0$ for any $\epsilon > 0$. Hence we conclude that $p(g) < 0$, and this implies $p(y) < 0$. Since $p(y) > 0$, we can say that $(x^2 + y^2 - 1)^3 > x^2 y^3$, as desired. Hence $x^2 + y^2 \le 2$ for any $(x, y) \in H$.

1.18 ▪ Generalized Assignment Problem

For part **(a)** define x_{ij} to be one if object i is stored in warehouse j. Then we have the model

$$
\begin{aligned}
\text{minimize} \quad & \sum_{i=1}^{10} \sum_{j=1}^{3} d_j x_{ij} \\
\text{subject to} \quad & \sum_{j=1}^{3} x_{ij} = 1, \quad \forall i \in \{1, \dots, 10\}, \quad \text{(every object needs to be stored)} \\
& \sum_{i=1}^{10} c_{ij} x_{ij} \le b, \quad \forall j \in \{1, 2, 3\}, \quad \text{(capacity of warehouses)} \\
& X \in \mathbb{B}^{10 \times 3},
\end{aligned}
$$

which is an instance of the *Generalized Assignment Problem*. For part **(b)**, the following constraint takes care of the requirement:

$$x_{11} + x_{21} \le 1 + \frac{x_{31} + x_{51}}{2}$$

since when the LHS is equal to 2, both x_{31} and x_{51} should be equal to one. If the LHS is less than 2 (equal to 0 or 1), then the constraint is redundant.

1.19 ▪ Basketball Team Line-up

This problem is from [25].[2]

Define x_i to be 1 if the player i is chosen for the starting team and 0 otherwise.

$$
\begin{aligned}
\text{maximize} \quad & 3x_1 + 2x_2 + 2x_3 + x_4 + 3x_5 + 3x_6 + x_7 \\
\text{subject to} \quad & x_1 + x_3 + x_5 + x_7 \geq 3, \\
& x_3 + x_4 + x_5 + x_6 + x_7 \geq 2, \\
& x_2 + x_4 + x_6 \geq 1, \\
& 3x_1 + 2x_2 + 2x_3 + x_4 + 3x_5 + 3x_6 + 3x_7 \geq 10, \\
& 3x_1 + x_2 + 3x_3 + 3x_4 + 3x_5 + x_6 + 2x_7 \geq 10, \\
& x_1 + 3x_2 + 2x_3 + 3x_4 + 3x_5 + 2x_6 + 2x_7 \geq 10, \\
& 1 - x_6 \geq x_3, \\
& x_4 + x_5 \geq 2x_1, \\
& x_2 + x_3 \geq 1, \\
& x_1 + x_2 + x_3 + x_4 + x_5 + x_6 + x_7 = 5, \\
& \mathbf{x} \in \mathbb{B}^7.
\end{aligned}
$$

1.20 ▪ Non-linearities

For part **(a)** we have

$$
\begin{aligned}
\text{minimize} \quad & 2x + 3y \\
\text{subject to} \quad & x + y \leq 3, \\
& x - y \leq 3, \\
& -x + y \leq 3, \\
& -x - y \leq 3, \\
& x, y \in \mathbb{Z}.
\end{aligned} \qquad (P_1)
$$

For part **(b)** of the problem we have

$$
\begin{aligned}
\text{minimize} \quad & 2x + 3y \\
\text{subject to} \quad & u + v \leq 3, \\
& -u \leq x \leq u, \\
& -v \leq y \leq v, \\
& x, y, u, v \in \mathbb{Z}.
\end{aligned} \qquad (P_2)
$$

---

[2]Used with permission of Cengage Learning.

1.21 ▪ Product of Two Binary Variables

This problem is from [20].[3]

Clearly, the $\delta_1\delta_2 = \delta_3$ requirement is equivalent to $[\delta_3 = 1 \Leftrightarrow \delta_1 = 1 \text{ and } \delta_2 = 1]$, which is equivalent to $[\delta_3 = 1 \Leftrightarrow \delta_1 + \delta_2 = 2]$. We can split this equivalence into two implications:

$$\delta_3 = 1 \implies \delta_1 + \delta_2 \geq 2 \quad \text{and} \quad \delta_1 + \delta_2 = 2 \implies \delta_3 = 1.$$

The following two inequalities achieve the desired result:

$$\delta_1 + \delta_2 \geq 2\delta_3,$$
$$\delta_1 + \delta_2 \leq 1 + \delta_3.$$

For the first inequality, if $\delta_3 = 1$, then indeed both δ_1 and δ_2 should be equal to one. If $\delta_3 = 0$, then the inequality says $\delta_1 + \delta_2 \geq 0$, which is a redundant piece of information. For the second inequality, if both δ_1 and δ_2 are equal to one, then δ_3 has to take value one. In case the sum of δ_1 and δ_2 is less than two, the inequality is redundant.

1.22 ▪ Sorting

Let Z be a binary matrix. Then the following is an equivalent mixed integer linear program:

$$
\begin{aligned}
\text{maximize} \quad & \mathbf{c}^T\mathbf{x} + \mathbf{f}^T\mathbf{y} \\
\text{subject to} \quad & A\mathbf{x} \leq \mathbf{b}, \\
& -M \leq x_i \leq M, & & \forall i \in \{1,\ldots,N\}, \\
& \sum_{i=1}^{N} z_{ij} = 1, & & \forall j \in \{1,\ldots,N\}, \\
& \sum_{j=1}^{N} z_{ij} = 1, & & \forall i \in \{1,\ldots,N\}, \\
& -M(1 - z_{ij}) \leq y_j - x_i, & & \forall i,j \in \{1,\ldots,N\}, \\
& y_j - x_i \leq M(1 - z_{ij}), & & \forall i,j \in \{1,\ldots,N\}, \\
& y_i \leq y_{i+1}, & & \forall i \in \{1,\ldots,N-1\}, \\
& Z \in \mathbb{B}^{N \times N}, \\
& \mathbf{x}, \mathbf{y} \in \mathbb{R}^{N}.
\end{aligned}
\tag{$\mathcal{P}$}
$$

---

[3]Reproduced by permission of Taylor and Francis Group, LLC, a division of Informa PLC.

1.23 ▪ Magic Squares

Let $x_{ijk} = 1$ if there is number k in row i and column j, and let t be the magic constant.

$$
\begin{aligned}
\text{maximize} \quad & t \\
\text{subject to} \quad & \sum_{i=1}^{n} \sum_{k=1}^{n^2} k x_{ijk} = t, \qquad \forall j \in \{1, \ldots, n\}, \\
& \sum_{j=1}^{n} \sum_{k=1}^{n^2} k x_{ijk} = t, \qquad \forall i \in \{1, \ldots, n\}, \\
& \sum_{i=1}^{n} \sum_{k=1}^{n^2} k x_{iik} = t, \\
& \sum_{i=1}^{n} \sum_{k=1}^{n^2} k x_{i(n+1-i)k} = t, \\
& \sum_{i=1}^{n} \sum_{j=1}^{n} x_{ijk} = 1, \qquad \forall k \in \{1, \ldots, n^2\}, \\
& \sum_{k=1}^{n^2} x_{ijk} = 1, \qquad \forall i, j \in \{1, \ldots, n\}, \\
& X \in \mathbb{B}^{n \times n \times n^2}, \\
& t \in \mathbb{R}.
\end{aligned}
$$

1.24 ▪ Lights-out Puzzle or the Game of Fiver

This problem is borrowed from the lecture notes of an MIT optimization course "Introduction to Integer Programming - MIT OpenCourseWare."[4]

Assume that each cell of the board has coordinates i, j, where $i, j \in \{1, \ldots, 5\}$. Let $x_{ij} = 1$ if cell (i, j) is clicked. The key to solving the puzzle is to realize that for all cells to be off at the end of the game each cell must undergo an odd number of status changes, i.e., one, three, five, seven, etc. Hence we have the following integer optimization formulation:

$$
\begin{aligned}
\text{minimize} \quad & \sum_{i=1}^{5} \sum_{j=1}^{5} x_{ij} \\
\text{subject to} \quad & x_{i-1,j} + x_{i+1,j} + x_{i,j-1} + x_{i,j+1} + x_{ij} = 2z_{ij} + 1, \quad \forall i, j \in \{1, \ldots, 5\}, \\
& X \in \mathbb{B}^{5 \times 5}, \\
& Z \in \mathbb{Z}_{+}^{5 \times 5}.
\end{aligned}
$$

The non-negative integer variables z_{ij} serve to count the odd number of changes to cell (i, j). The reader who wishes to try out the puzzle can consult the following web site:

https://www.bsswebsite.me.uk/Puzzlewebsite/Lightsoutpuzzle/lightsout.html

---

[4]https://ocw.mit.edu/courses/sloan-school-of-management/15-053-optimization-methods-in-managem
ent-science-spring-2013/lecture-notes/MIT15_053S13_lec10.pdf

1.25 ▪ Sparse Coding

The troubling aspect of the above formulation is the constraint that limits the number of non-zero elements of the solution to T. To handle this issue, we introduce binary variables z_i for each continuous variable x_i and relate the two variables by the constraints

$$-Mz_i \leq x_i \leq Mz_i, \quad \forall i \in \{1,\ldots,p\},$$

where M is a suitably large constant. These constraints make sure that $z_i = 1$ if and only if $x_i \neq 0$. Then we obtain the following mixed integer quadratic programming problem:

$$
\begin{aligned}
\text{minimize} \quad & \|y - Dx\|_2^2 \\
\text{subject to} \quad & -Mz_i \leq x_i \leq Mz_i, \quad \forall i \in \{1,\ldots,p\}, \\
& \sum_{i=1}^{p} z_i \leq T, \\
& \mathbf{x} \in \mathbb{R}^p, \\
& \mathbf{z} \in \mathbb{B}^p.
\end{aligned}
$$

1.26 ▪ Partition to Eliminate Triangles

This problem is from the MATP6620/ISYE6760 Combinatorial Optimization & Integer Programming course by J.W. Mitchell.[5]

Define the binary variable x_e to indicate whether an edge is in E_1:

$$
x_e = \begin{cases} 1 & \text{if } e \in E_1, \\ 0 & \text{if } e \in E_2. \end{cases}
$$

To eliminate triangles, we must reason as follows. If three edges constitute a triangle in E, then we need the sum of their x-values to be equal to either one or two. Otherwise, if the sum is equal to three, then all three edges are in E_1, and if the sum is nil, then all three edges are in E_2. So the binary variables x_e must satisfy the constraints

$$1 \leq x_e + x_f + x_g \leq 2, \text{ for all triangles } e, f, g \in E.$$

1.27 ▪ A Production Planning Problem

This problem is adapted from the IE418 Integer Programming course slides of Prof. Jeff Linderoth, University of Wisconsin-Madison.

---

[5]https://homepages.rpi.edu/~mitchj/matp6620

For part **(a)**, let x_i, where $i \in \{1, \ldots, 5\}$, denote the *integer non-negative* variables representing the number of product i to be manufactured during the week.

$$\text{maximize} \quad 55x_1 + 60x_2 + 35x_3 + 40x_4 + 20x_5$$

$$\text{subject to} \quad 12x_1 + 20x_2 + 25x_4 + 15x_5 \leq 288, \quad \text{(grinding constraint)}$$
$$10x_1 + 8x_2 + 16x_3 \leq 192, \quad \text{(drilling constraint)}$$
$$20(x_1 + x_2 + x_3 + x_4 + x_5) \leq 384. \quad \text{(final assembly constraint)}$$

For part **(b)**, we first define an indicator variable $z_i \in \mathbb{B}$ to indicate whether or not $x_i > 0$. We want the implication

$$x_i > 0 \iff z_i = 1$$

to hold. We define the range constraint

$$z_i \leq x_i \leq M z_i,$$

where a positive sufficiently large constant M can be deduced from the resource constraints, e.g., for p_1, M can be chosen as

$$\min \left\{ \left\lfloor \frac{288}{12} \right\rfloor, \left\lfloor \frac{192}{10} \right\rfloor, \left\lfloor \frac{384}{20} \right\rfloor \right\} = 19.$$

One repeats this construction for all variables x_i. Now, having defined the indicator variables z_i, we need to model the implication

$$z_1 + z_2 \geq 1 \implies z_3 + z_4 + z_5 \geq 1.$$

We shall do this in two steps: we shall define another indicator variable $\delta \in \mathbb{B}$ and model the two-step implication:

$$z_1 + z_2 \geq 1 \implies \delta = 1,$$
$$\delta = 1 \implies z_3 + z_4 + z_5 \geq 1.$$

First, we deal with

$$z_1 + z_2 \geq 1 \implies \delta = 1.$$

The inequality

$$z_1 + z_2 \leq 2\delta$$

does what we want since if $\delta = 0$, both $z_1 = z_2 = 0$. If $\delta = 1$, the inequality is redundant. Now, for the second part we need to make sure

$$\delta = 1 \implies z_3 + z_4 + z_5 \geq 1.$$

The inequality

$$z_3 + z_4 + z_5 \geq \delta$$

is what we need. Indeed, if $\delta = 1$, at least one of p_3, p_4, p_5 will be produced. If $\delta = 0$, the inequality is redundant.

1.28 ▪ Telecommunications Network Design

This problem is adapted from [9].

The network has nodes $\mathcal{N} = \{1, \ldots, M\} \cup \{i^*\}$, i.e., the ring nodes and the MTSO. Define binary variables x_{cj}, which is equal to one if cell c is connected to node j and equal to zero otherwise. Then we have the objective function

$$
\begin{aligned}
\text{minimize} \quad & \sum_{c \in \mathcal{C}} \sum_{j \in \mathcal{N}} co_{cj} x_{cj} \\
\text{subject to} \quad & \sum_{j \in \mathcal{N}} x_{cj} = d_c, & \forall c \in \mathcal{C}, \quad (I) \\
& \sum_{c \in \mathcal{C}} \sum_{j \in \mathcal{N} \setminus \{i^*\}}^{M} \tfrac{T_c}{d_c} x_{cj} \le 2 \cdot U, & (II) \\
& X \in \mathbb{B}^{|\mathcal{C}| \times |\mathcal{N}|}.
\end{aligned}
$$

The number of connections for each cell c must be equal to the diversity of the cell, expressed by constraint (I).

A necessary condition for keeping within the capacity limits of the ring is that all demands of any origin are routed through the MTSO, following the ring in one direction or the other. Since every edge of the ring has a capacity U, the total traffic on the ring may not exceed $2 \cdot U$. Note that the traffic of a cell directly connected to the MTSO does not enter the ring. This constraint is expressed in (II).

1.29 ▪ Combinatorial Auctions: Winner Determination

This and the following two problems are inspired from the lecture notes of CPS296 by V. Conitzer, Duke University.[6]

For every bid $b \in \mathcal{B}$, where $\mathcal{B}$ is the set of bids, let there be a variable x_b that takes values in $\{0, 1\}$; setting this variable to 1 means that the bid is accepted, and setting it to 0 means that it is rejected. For each bid b and item i, let there be a parameter a_{ib} which is 1 if item i is included in bid b, and 0 if it is not. Finally, let v_b be the value of bid b. Then we can write the winner determination problem as follows:

$$
\begin{aligned}
\text{maximize} \quad & \sum_{b \in \mathcal{B}} v_b x_b \\
\text{subject to} \quad & \sum_{b \in \mathcal{B}} a_{ib} x_b \le 1, \quad \forall i \in I, \\
& \mathbf{x} \in \mathbb{B}^{|\mathcal{B}|}.
\end{aligned}
$$

---

[6]https://courses.cs.duke.edu/fall10/cps296.1/applications.pdf

1.30 ▪ Combinatorial Reverse Auctions

Let $\mathcal{B}$ be the set of bids. We can model this as an integer program as follows. Again x_b is a binary variable equal to one if bid $b \in \mathcal{B}$ is accepted, and equal to 0 otherwise.

$$
\begin{aligned}
\text{minimize} \quad & \sum_{b \in \mathcal{B}} v_b x_b \\
\text{subject to} \quad & \sum_{b \in \mathcal{B}} a_{ib} x_b \geq 1, \quad \forall i \in I, \\
& \mathbf{x} \in \mathbb{B}^{|\mathcal{B}|},
\end{aligned}
$$

where, again, a_{ib} is a parameter indicating whether item i occurs in bid b.

1.31 ▪ Rank Aggregation: Kemeny Rule

For every ordered pair of alternatives a, b, let x_{ab} be a binary variable that is 1 if a is ranked ahead of b in the aggregate ranking and 0 otherwise. We need

$$
x_{ab} + x_{ba} = 1
$$

(exactly one of a and b must be ahead of the other). We must also have

$$
x_{ab} + x_{bc} + x_{ca} \leq 2,
$$

because if all of these variables are equal to 1, then that means that a is ranked ahead of b, b is ranked ahead of c, and c is ranked ahead of a, which is impossible. Conversely, if the variables are set in such a way that these constraints hold, then this will in fact correspond to a ranking. Now, if we let n_{ab} (a parameter) be the number of input rankings that rank a ahead of b, then we seek to minimize

$$
\sum_{a \neq b} n_{ba} x_{ab}.
$$

Putting it all together, we obtain

$$
\begin{aligned}
\text{minimize} \quad & \sum_{a \neq b} n_{ba} x_{ab} \\
\text{subject to} \quad & x_{ab} + x_{ba} = 1, \quad & \forall a, b \quad & \text{such that} \quad a \neq b, \\
& x_{ab} + x_{bc} + x_{ca} \leq 2, \quad & \forall a, b, c \quad & \text{such that} \quad a \neq b, b \neq c, c \neq a.
\end{aligned}
$$

1.32 ▪ Coding Theory and Integer Optimization

This problem is based on Helmling et al. [10].

The only thing to do is to write the equations of the system

$$H\mathbf{x} = \mathbf{0} \mod 2.$$

For row i, this is achieved by the constraint

$$(H\mathbf{x})_i = 2 * z_i,$$

where z_i is a non-negative integer variable for each row i of the matrix H.

1.33 ▪ Truss Topology Design by Mixed Integer Optimization

This problem is adapted from Kocvara [12].

One has to add the following constraints:

$$0 \le t_i \le T \quad \text{and} \quad t_i \in \mathbb{Z}, \quad \text{for all } i.$$

1.34 ▪ Sequence Alignment in Computational Biology

This problem is adapted from Lancia [13].

Observe that a non-crossing matching in W_{nm} corresponds to a stable set (or an independent set) in the graph G_L. Since a stable set can intersect a clique in at most 1 node, the given clique constraints can be used as constraints in the alignment problem.

1.35 ▪ Election Rigging or Gerrymandering

This problem requires a pre-processing step where all potential districts are formed based on the criteria listed in the description. We assume that this has been performed and given, as well as P_d values.

We define a binary variable x_d per possible district d that takes the value 1 if the district is chosen for the partitioning. Then we have the objective function

$$
\begin{aligned}
\text{maximize} \quad & \sum_{d \in R} P_d x_d \\
\text{subject to} \quad & \sum_{d \in R} x_d = q, & (I) \\
& \sum_{d \in R} Q_{di} x_d = 1, \quad \forall i \in \{1, \dots, M\}. & (II)
\end{aligned}
$$

Constraint (I) forces that exactly q districts are formed and (II) forces that every county i has to appear in exactly one district of the chosen partitioning.

1.36 ▪ Traveling Salesperson with Due Dates

Recall the Miller–Tucker–Zemlin constraints for the asymmetric TSP with n nodes in the graph:

$$u_i - u_j + n x_{ij} \leq n - 1,$$

where u_i represents the sequence of visit to node i ($u_1 = 1$).

We start the formulation by defining the binary variables x_{ij} as equal to one if city j is visited immediately after city i, and equal to zero otherwise. Then we have the following model:

$$
\begin{aligned}
\text{minimize} \quad & \sum_{(i,j) \in A} t_{ij} x_{ij} \\
\text{subject to} \quad & \sum_{(i,j):(i,j) \in A} x_{ij} = 1, & \forall i \in N, \\
& \sum_{(k,i):(k,i) \in A} x_{ki} = 1, & \forall i \in N, \\
& u_j \geq u_i + t_{ij} + M(x_{ij} - 1), & \forall (i,j) \in A, i \neq j, j \neq 1, \\
& u_i \leq d_i, & \forall i \in N \setminus \{1\}, \\
& u_1 = 0,
\end{aligned}
$$

where the last three constraints are the modified Miller–Tucker–Zemlin (MMTZ) constraints. Note that the MMTZ constraints eliminate all potential cycles since for any cycle $C \subset N$ and the arcs of the cycle $E(C) = \{(i,j) \in A : i \in C, j \in C\}$, summing up the inequalities MMTZ for all arcs in $E(C)$ we arrive at the inequality

$$0 \geq \sum_{(i,j) \in E(C)} t_{ij},$$

which is impossible.

1.37 ▪ Prize Collecting Salesperson

In addition to the usual binary variables x_{ij} defined as in the previous problem, we define variable $y_i = 1$ if the salesperson visits city i and zero otherwise and assume that node 1 is the depot. Then we have the assignment constraints (modified) and the usual Miller–Tucker–Zemlin constraints in addition to prize collection to obtain the following formulation:

$$\text{minimize} \quad \sum_{(i,j)\in A} t_{ij}x_{ij} + \sum_{i\in N} c_i(1 - y_i)$$

$$\text{subject to} \quad \sum_{(i,j):(i,j)\in A} x_{ij} = y_i, \qquad \forall i \in N,$$

$$\sum_{(k,i):(k,i)\in A} x_{ki} = y_i, \qquad \forall i \in N,$$

$$u_i - u_j + nx_{ij} \leq n - 1, \qquad \forall (i,j) \in A, i \neq j, i, j \neq 1,$$

$$\sum_{i\in N} w_i y_i \geq W,$$

$$u_i \geq 0, \qquad \forall i \in N \setminus \{1\},$$

$$y_1 = 1,$$

$$u_1 = 1,$$

where u_i represents the sequence of visit to node i.

1.38 ▪ Multiple Salespersons

We define the usual binary variables x_{ij} as equal to 1 if city j is visited immediately after city i. Then the following formulation is obtained with the modified Miller–Tucker–Zemlin (MMTZ) constraints:

$$\text{minimize} \quad \min \sum_{(i,j)\in A} c_{ij}x_{ij}$$

$$\text{subject to} \quad \sum_{(i,j):(i,j)\in A} x_{ij} = 1, \qquad \forall i \in N \setminus \{1\},$$

$$\sum_{(k,i):(k,i)\in A} x_{ki} = 1, \qquad \forall i \in N \setminus \{1\},$$

$$\sum_{(1,j)\in A} x_{1j} = m,$$

$$\sum_{(k,1)\in A} x_{k1} = m,$$

$$u_j \geq u_i + 1 + (n - m)(x_{ij} - 1), \quad \forall (i,j) \in A, i \neq j, j \neq 1,$$

$$u_i \geq 0, \qquad \forall i \in N \setminus \{1\},$$

$$u_1 = 0.$$

1.39 ▪ Clustering in Multivariate Data Analysis

Based on the considerations given above we have the following formulation:

$$\text{minimize} \quad \sum_{i=1}^{n}\sum_{j=1}^{n} d_{ij}x_{ij}$$

$$\text{subject to} \quad \sum_{i=1}^{n} x_{ij} = 1, \qquad \forall j \in \{1,\ldots,n\},$$

$$\sum_{j=1}^{n} x_{jj} = m,$$

$$\sum_{i=1}^{n} x_{ij} \leq m_0 x_{jj}, \quad \forall j \in \{1,\ldots,n\}.$$

In the formulation, the variables x_{jj} refer to the choice of element j as the median of a cluster.

1.40 ▪ Support Vector Machines. Feature Selection Budget

This problem is adapted from [14].

Define a binary variable v_j equal to one if feature j is selected (assuming for convenience that weight vectors w are now bounded by the interval $[-1, 1]$):

$$
\begin{aligned}
\text{minimize} \quad & \sum_{i=1}^{m} \xi_i \\
\text{subject to} \quad & y_i(w^T x_i + b) \geq 1 - \xi_i, \quad \forall i \in \{1, \ldots, m\}, \\
& -v_j \leq w_j \leq v_j, \quad \forall j \in \{1, \ldots, n\}, \\
& \sum_{j=1}^{n} c_j v_j \leq B, \\
& \boldsymbol{\xi} \in \mathbb{R}_{+}^{m}, \\
& \mathbf{v} \in \mathbb{B}^n.
\end{aligned}
$$

1.41 ▪ Network Revenue Management

The reader may consult Chapter 8 of Philips [16] for background on this problem.

Define integer-valued non-negative variables x_{pj}, where p stands for the city origin-destination pairs from $\{CB, K, W\}$ and $j \in \{Y, M, Q\}$ designates different fare classes. For instance, x_{CBKY} represents the number of seats in fare class Y reserved for passengers flying from Congo-Brazzaville to Kampala.

(a) The formulation is as follows:

$$
\begin{aligned}
\text{maximize} \quad & 500x_{CBKY} + 300x_{CBKM} + 150x_{CBKQ} + \cdots + 250x_{CBWQ} \\
\text{subject to} \quad & x_{CBKY} + x_{CBKM} + x_{CBKQ} + x_{CBWY} + x_{CBWM} + x_{CBWQ} \\
& \leq 100, \\
& x_{KWY} + x_{KWM} + x_{KWQ} + x_{CBWY} + x_{CBWM} + x_{CBWQ} \\
& \leq 100, \\
& x_{CBKY} \leq 10, \\
& x_{CBKm} \leq 20, \\
& \vdots \\
& x_{CBWQ} \leq 45.
\end{aligned}
$$

(b) The optimal allocations are found as follows using the two constraints above as tight at optimality: $x_{CBKY} = 10$, $x_{CBKM} = 20$, $x_{CBWY} = 8$, $x_{CBWM} = 25$, $x_{KWY} = 15$, $x_{KWM} = 25$, $x_{KWQ} = 27$.

1.42 ▪ Shirley's Birthday

(a) Define the decision variable

$$x_{ij} = \begin{cases} 1 & \text{if friend } i \text{ gives coin type } j, \\ 0 & \text{otherwise.} \end{cases}$$

Add the matching like constraints; each friend will give one type of coin:

$$\text{maximize} \quad \sum_{i=1}^{n} \sum_{j=1}^{3} x_{ij} c_{ij}$$

$$\text{subject to} \quad \sum_{j=1}^{3} x_{ij} = 1, \quad \forall i \in \{1 \ldots n\},$$

$$X \in \mathbb{B}^{n \times 3}.$$

(b) Adding the following inequalities suffices:

$$\sum_{i=1}^{n} x_{i1} \leq 30,$$

$$\sum_{i=1}^{n} x_{i3} \geq 40.$$

Chapter 2

Shortest Paths, Maximal Flows, and Trees

2.1 ▪ Shortest Paths with Two Sources and Two Sinks

Add two extra nodes $\{S^*, T^*\}$ and create arcs $(S^*, S_1), (S^*, S_2), (T_1, T^*), (T_2, T^*)$ with zero cost. Then use some algorithm to solve the shortest path. An optimal solution is $S_2 \rightarrow 2 \rightarrow 5 \rightarrow T_1$ with length 10.

2.2 ▪ Shortest Path with Restrictions

(a) Let x_{ij} be the binary variable with value 1 if arc (i, j) is in the shortest path. Let $\delta^+(i) = \{j : (i, j) \in E\}$ and $\delta^-(i) = \{k : (k, i) \in E\}$. Then we have the formulation

$$
\begin{aligned}
\text{minimize} \quad & \sum_{(i,j) \in E} x_{ij} \\
\text{subject to} \quad & \sum_{j \in \delta^+(s)} x_{sj} = 1, \\
& \sum_{j \in \delta^-(t)} x_{jt} = 1, \\
& \sum_{j \in \delta^+(i)} x_{ij} = \sum_{k \in \delta^-(i)} x_{ki}, \quad \forall i \in N \setminus \{s, t\}, \\
& \sum_{(i,j) \in E} c_{ij} x_{ij} \leq M, \\
& x_{ij} \in \mathbb{B}, \quad\quad\quad\quad\quad \forall (i, j) \in E.
\end{aligned}
$$

(b) In this case, we have the usual shortest path formulation with an additional constraint $(*)$,

$$
\begin{aligned}
\text{minimize} \quad & \sum_{(i,j)\in E} c_{ij}x_{ij} \\
\text{subject to} \quad & \sum_{j\in\delta^+(s)} x_{sj} = 1, \\
& \sum_{j\in\delta^-(t)} x_{jt} = 1, \\
& \sum_{j\in\delta^+(i)} x_{ij} = \sum_{k\in\delta^-(i)} x_{ki}, \quad \forall i \in N \setminus \{s,t\}, \\
& \sum_{(i,j)\in E} c_{ij}x_{ij} \leq M, \\
& \sum_{(i,j)\in E} x_{ij} \leq p+1, \qquad\qquad\qquad\qquad (*) \\
& x_{ij} \in \mathbb{B}, \qquad\qquad\qquad \forall(i,j) \in E.
\end{aligned}
$$

2.3 ▪ Shortest Path Duality

This problem is taken from Young.[7]

(a) Using the Dijkstra algorithm or simply by inspection, this problem returns the path $s - 3 - 6 - t$ with distance 10.

(b) We can formulate the problem as

$$
\begin{aligned}
\text{maximize} \quad & u_t - u_s \\
\text{subject to} \quad & u_j - u_i \leq d_{ij}, \quad \forall(i,j) \in E, \\
& \mathbf{u} \in \mathbb{B}^{|V|}.
\end{aligned}
$$

(c) Formulations for parts (a) and (b) are primal-dual pairs. Since you found the optimal solution for part (a), using strong duality we say that optimal value for part (b) is 10.

2.4 ▪ Constrained Shortest Path and Saving Corporal Raymond

This problem is adapted from Fischetti [7].

Given a directed complete graph $G = (V, E)$ with positive arc costs c_{ij}, two distinct vertices v_1 (the command line) and v_2 (Corporal Raymond), and a subset $Q \in V \setminus \{v_1, v_2\}$ of vertices (safe posts), the problem proposed is to find a shortest path from v_1 to v_2 that visits at least half of the nodes in Q. Define the binary decision variables x_{ij} for each arc $(i, j) \in E$, equal to one if arc (i, j) is selected. We have the following

---

[7]http://www.cs.ucr.edu/~neal/1998/cosc185-S98/linear-programming/examples/node4.html#:~: text=representing%20the%20position%20of%20the,the%20line%20than%20vertex%20u

formulation with the path defining constraints and contraint $(*)$ that at least half the nodes in Q should be visited:

$$\begin{array}{ll}
\text{minimize} & \sum_{i \in V} \sum_{j \in V} c_{ij} x_{ij} \\[2mm]
\text{subject to} & \sum_{j \in V} x_{v_1 j} = 1, \\[2mm]
& \sum_{j \in V} x_{j v_2} = 1, \\[2mm]
& \sum_{j \in V} x_{ij} = \sum_{k \in V} x_{ki}, \qquad \forall i \in V \setminus \{v_1, v_2\}, \\[2mm]
& \sum_{h \in Q} \sum_{i \in V} x_{ih} \geq \left\lceil \frac{|Q|}{2} \right\rceil, \qquad\qquad\qquad (*) \\[2mm]
& x_{ij} \in \mathbb{B}, \qquad\qquad\qquad \forall (i,j) \in E.
\end{array}$$

2.5 ▪ LP Relaxation for Matching

(a) Recall that in the maximum cardinality matching problem we define the binary variables x_e or each edge in the graph. This variable takes value one if edge e is in the matching. Adding all the matching constraints

$$\sum_{e = (i,j) \text{ or } (j,i)} x_e \leq 1, \quad \forall i \in V,$$

one obtains the inequality

$$\sum_{e \in E} x_e \leq \frac{9}{2}.$$

This is an upper bound for all feasible solutions of the LP relaxation since we are maximizing $\sum_{e \in E} x_e$. Take the 9 edges (A, B), (B, C), (C, G), (G, H), (H, J), (D, J), (D, E), (E, F), and (A, F), and put their corresponding variables to $\frac{1}{2}$. This feasible solution attains the upper bound. Hence, it is optimal for the LP relaxation.

(b) The matching number of the graph is equal to 4; take, e.g., (A, B), (C, G), (D, J), and (E, F) in the matching.

(c) Recall that odd-cycle elimination inequalities are valid inequalities in the matching problem. It suffices to add

$$\sum_{e \in E} x_e \leq 4$$

to the LP relaxation to get rid of the fractional LP solution.

2.6 ▪ LP Relaxation for Edge Covering

(a) The minimum cardinality edge covering problem by nodes is formulated by defining a binary variable y_i for each node i in the graph as equal to one if node i is in
the cover. We have the following integer optimization model:

$$\begin{aligned}
\text{minimize} \quad & \sum_{i \in V} y_i \\
\text{subject to} \quad & y_i + y_j \geq 1, \quad \forall (i,j) \in E, \\
& \mathbf{y} \in \mathbb{B}^{|V|}.
\end{aligned}$$

The LP relaxation is obtained by simply ignoring the binary nature of variables
and imposing a non-negativity restriction on all y_is (as we are minimizing, no
variable will take a value larger than one). Now, we notice that the LP relaxation

$$\begin{aligned}
\text{minimize} \quad & \sum_{i \in V} y_i \\
\text{subject to} \quad & y_i + y_j \geq 1, \quad \forall (i,j) \in E, \\
& \mathbf{y} \in \mathbb{R}_+^{|V|}
\end{aligned}$$

is dual to the LP relaxation of maximal cardinality matching:

$$\begin{aligned}
\text{maximize} \quad & \sum_{e \in E} x_e \\
\text{subject to} \quad & \sum_{e=(i,j) \text{ or } (j,i)} x_e \leq 1, \quad \forall i \in V, \\
& \mathbf{x} \in \mathbb{R}_+^{|E|}.
\end{aligned}$$

Since both primal and dual LPs are feasible, they are both solvable and have equal
optimal values. Since we have found in Problem 2.5 that the optimal value of the
matching LP is equal to 4.5, the optimal value of the covering LP must also be
equal to 4.5. This is indeed true since we can take all y_i variables to be equal to
$\frac{1}{2}$ and get a feasible solution with value equal to 4.5.

(b) By inspection, the covering number of the graph is found to be equal to 6.

2.7 ▪ Pairing Surgeons

This problem is modified from [9].

This problem calls for a maximum cardinality matching formulation on an undirected graph $G = (V, E)$ formed as follows. For each surgeon we form a node. We
make an edge between two surgeons i and j if each surgeon has at least $\frac{11}{20}$ for the same

language and $\frac{11}{20}$ on the same specialization. Then we solve the following problem:

$$\text{maximize} \quad \sum_{(i,j)\in E} x_{ij}$$

$$\text{subject to} \quad \sum_{(i,j)\in\delta(i)} x_{ij} \leq 1,$$

$$\mathbf{x} \in \mathbb{B}^{|E|},$$

where $\delta(i) = \{(i,j) \text{ or } (k,i) : (i,j), (k,i) \in E\}$ is the set of all edges in E incident to node i.

2.8 ▪ Steiner Trees

Define the binary variables x_{ij} for each arc $(i,j) \in A$ as equal to one if arc (i,j) is selected and as zero otherwise. We also define two sets $\Delta^-(j) = \{(i,j) \in E\}$ and $\Delta^+(S) = \{(i,j) \in E : i \in S, j \notin S\}$. Then we have the following formulation:

$$\text{minimize} \quad \sum_{(i,j)\in A} c_{ij} x_{ij}$$

$$\text{subject to} \quad \sum_{(i,r)\in\Delta^-(r)} x_{ij} = 0,$$

$$\sum_{(i,j)\in\Delta^-(j)} x_{ij} = 1, \qquad \forall j \in T,$$

$$\sum_{(i,j)\in\Delta^+(S)} x_{ij} \geq \sum_{(i,t)\in\Delta^-(t)} x_{it}, \quad \forall S \subset V, \text{ where } r \in S, \text{ and } t \in V \setminus S,$$

$$x_{ij} \in \mathbb{B}, \qquad \forall (i,j) \in E.$$

In degree constraints, the LHS represents the number of arcs entering a node: the root r should have no entering arc, the destination nodes in T exactly one entering arc, and the remaining nodes 0 or 1 entering arc.

The cut constraints say that every non-isolated vertex t (i.e., having exactly one entering arc) should be attainable from r: every cut $\Delta^+(S)$ with $r \in S$ and $t \notin S$ should contain at least one selected arc.

2.9 ▪ Maximum Weight Spanning Trees

Define a binary variable x_e for each edge e, which takes value one if e is in the spanning tree. Then, the objective function to be maximized is

$$\max \sum_{e\in E} c_e x_e.$$

Using the hint, we have the constraint

$$\sum_{e\in E} x_e = |V| - 1.$$

We also need to prevent cycles: for each cycle $C \subseteq V$, define $E(C) = \{e = (i, j) \in E : i \in C$ and $j \in C\}$ as the set of all edges of the cycle C; then we write the inequality

$$\sum_{e \in E(C)} x_e \leq |C| - 1,$$

which is to be written for each cycle of cardinality greater than or equal to three in the graph. Since we are maximizing and edge weights are positive, we can relax the equality in the first constraint to a less-than-or-equal type inequality. Hence we have the formulation for maximum weight spanning tree

$$\begin{aligned}
\text{maximize} \quad & \sum_{e \in E} c_e x_e \\
\text{subject to} \quad & \sum_{e \in E} x_e \leq |V| - 1, \\
& \sum_{e \in E(C)} x_e \leq |C| - 1, \quad \forall C \subseteq V, |C| \geq 3, \\
& \mathbf{x} \in \mathbb{B}^{|E|}.
\end{aligned}$$

If there is a cycle containing all vertices, the relaxed equality constraint will be implied by the cycle constraint, and we will be dealing with a smaller optimization problem.

2.10 ▪ Minimum Weight Spanning Trees

Students are usually tempted to take the formulation for maximum weight spanning trees and write "min" instead of "max" in the objective function and walk away:

$$\begin{aligned}
\text{minimize} \quad & \sum_{e \in E} c_e x_e \\
\text{subject to} \quad & \sum_{e \in E} x_e \leq |V| - 1, \\
& \sum_{e \in E(C)} x_e \leq |C| - 1, \quad \forall C \subseteq V, |C| \geq 3, \\
& \mathbf{x} \in \mathbb{B}^{|E|}.
\end{aligned}$$

However, this formulation is useless for the minimum weight spanning tree problem since it results in an optimal solution with weight equal to zero and no edge chosen in the spanning tree. However, the solution is obtained by switching the inequality in the first constraint into an equality.

2.11 ▪ Disasters and Spanning Trees

Assume that all houses are connected in a spanning tree structure T with edges $E(T)$. Under the independence of the events that there are roadblocks between i and j, the probability that no connection is compromised is equal to

$$\prod_{(i,j)\in E(T)} (1 - p_{ij}).$$

Hence, the probability that a route is blocked is equal to

$$1 - \prod_{(i,j)\in E(T)} (1 - p_{ij}),$$

which is to be minimized. Therefore, we have to solve

$$\min_{T\in\mathcal{T}} 1 - \prod_{(i,j)\in E(T)} (1 - p_{ij}),$$

where $\mathcal{T}$ denotes the set of all spanning trees connecting all houses. The problem is equivalent to the following integer programming problem:

$$
\begin{aligned}
\text{maximize} \quad & \sum_{(i,j)\in E} \ln(1 - p_{ij})x_{ij} \\
\text{subject to} \quad & \sum_{e=(i,j)\in E} x_e \le |V| - 1, \\
& \sum_{e=(i,j)\in E(C)} x_e \le |C| - 1, \quad \forall C \subseteq V, |C| \ge 3, \\
& \mathbf{x} \in \mathbb{B}^{|E|}.
\end{aligned}
$$

2.12 ▪ Emergency Planning and Spanning Trees I

This problem is modified from Fischetti [7].

Let x_e be a binary variable, equal to one if edge e is selected. Then we have the formulation

$$
\begin{aligned}
\text{minimize} \quad & \sum_{e\in E} c_e x_e \\
\text{subject to} \quad & \sum_{e\in E} x_e = |V| - 1, \\
& \sum_{e\in E(C)} x_e \le |C| - 1, \quad \forall C \subseteq V, |C| \ge 3, \\
& \sum_{e\in E} w_e x_e \ge W, \\
& \sum_{e\in E_1} x_e \le \sum_{e\in E_2} x_e - 1, \\
& \mathbf{x} \in \mathbb{B}^{|E|}.
\end{aligned}
$$

2.13 ▪ Emergency Planning and Spanning Trees II

This problem is also adapted from Fischetti [7].

Let x_e be a binary variable, equal to one if edge e is selected. Then we have the formulation

$$
\begin{aligned}
\text{minimize} \quad & \sum_{e \in E} c_e x_e \\
\text{subject to} \quad & \sum_{e \in E} x_e = |V| - 1, \\
& \sum_{e \in E(C)} x_e \le |C| - 1, \quad \forall C \subseteq V, |C| \ge 3, \\
& \sum_{e \in E} w_e x_e \le W \sum_{e \in E} x_e, \\
& \sum_{e \in E_1} x_e \ge \left\lceil \tfrac{|V|-1}{2} \right\rceil, \\
& \mathbf{x} \in \mathbb{B}^{|E|}.
\end{aligned}
$$

2.14 ▪ Envy-Free Perfect Matching

Define a binary decision variable x_{bi}, equal to one if buyer b is assigned item i, and non-negative variables p_i for the price of CD i. Then we have the following formulation:

$$
\begin{aligned}
\text{maximize} \quad & \sum_{i=1}^{n} p_i \\
\text{subject to} \quad & \sum_{i=1}^{n} x_{bi} = 1, & \forall b \in \{1, \dots, n\}, \\
& \sum_{b=1}^{n} x_{bi} = 1, & \forall i \in \{1, \dots, n\}, \\
& v_{bi} x_{bi} - p_i \ge 0, & \forall i, b \in \{1, \dots, n\}, \\
& v_{bi} x_{bi} - p_i \ge v_{b\ell} x_{bi} - p_\ell, & \forall i, b \in \{1, \dots, n\}, \quad \forall \ell \ne i, \\
& \mathbf{p} \in \mathbb{R}_+^n, \\
& X \in \mathbb{B}^{n \times n}.
\end{aligned}
$$

2.15 ▪ Spies and B-Matchings

Let x_e be a binary variable equal to one if edge e is selected, and zero otherwise. Let the set of edges incident to node v be

$$
\delta(v) = \{e = (v, u) \text{ or } (w, v) \in E\}.
$$

Then we have the following formulation:

$$
\begin{aligned}
\text{minimize} \quad & \sum_{e \in E} c_e x_e \\
\text{subject to} \quad & \sum_{e \in \delta(v)} x_e = b_v, \quad \forall v \in V, \\
& \mathbf{x} \in \mathbb{B}^{|E|}.
\end{aligned}
$$

When $b_v = 2$ for all $v \in V$, we obtain the *2-matching* problem, which is a well-known relaxation for the Symmetric Traveling Salesperson Problem (STSP).

2.16 ▪ Matchings and Stable Marriages

This problem is adapted from [17].

Define the matching binary variable x_{mw} equal to one if man m and woman w are matched, and zero otherwise. The usual matching constraints are written first:

$$\sum_{j \in W} x_{mj} \leq 1, \quad \forall m \in M,$$

$$\sum_{t \in M} x_{tw} \leq 1, \quad \forall w \in W.$$

The individual rationality constraint is written as follows:

$$x_{mw} = 0, \quad \forall (m, w) \in M \times W \setminus A.$$

The stability constraint should ensure the following. If woman w chooses someone less desirable than m, then man m must marry someone more desirable than she. Together with the assignment constraints above the stability constraint also implies that if man m marries someone less desirable than woman w, then she must marry someone more desirable than he. This is expressed as

$$\sum_{j >_m w} x_{mj} + \sum_{t >_w m} x_{tw} + x_{mw} \geq 1, \quad \forall (m, w) \in A,$$

where the shorthand $\{j >_m w\}$ means the set $\{j \in W : j >_m w\}$ (the set of women m prefers to w), and the same for $\{t >_w m\}$.

2.17 ▪ Stable Marriage with Ties and Incomplete Lists

This problem is adapted from [6].

Define as usual binary decision variables x_{ij} that take value 1 if child i is matched with family j, and 0 otherwise ($i \in \{1, \ldots, n_1\}$, $j \in F(i)$). We have the matching constraints

$$\sum_{j \in F(i)} x_{ij} \leq 1, \quad i \in \{1, \ldots, n_1\},$$

$$\sum_{i \in C(j)} x_{ij} \leq 1, \quad j \in \{1, \ldots, n_2\}.$$

We need to define constraints that enforce stability by ruling out the existence of any blocking pair. More specifically, they must ensure that if child i is not matched with family j or any other family, they rank at the same level or better than j (i.e., $\sum_{q \in F_j^{\leq}(i)} x_{iq}$ $= 0$); then family j is matched with a child it ranks at the same level or better than i (i.e., $\sum_{p \in C_i^{\leq}(j)} x_{pj} \geq 1$). The following inequalities achieve this purpose:

$$1 - \sum_{q \in F_j^{\leq}(i)} x_{iq} \leq \sum_{p \in C_i^{\leq}(j)} x_{pj}, \quad \forall i \in \{1, \ldots, n_1\}, \quad \forall j \in F(i).$$

2.18 ▪ Hospitals and Residents with Ties

A variant of the previous problem concerns the assignment of medical doctors in training (resident doctors) to hospitals, and is adapted from [6].

Define as usual binary decision variables x_{ij} that take value 1 if doctor i is matched with hospital j, and 0 otherwise ($i \in \{1, \ldots, n_1\}$, $j \in H(i)$). We have the matching and hospital capacity constraints:

$$\sum_{j \in H(i)} x_{ij} \leq 1, \quad \forall i \in \{1, \ldots, n_1\},$$

$$\sum_{i \in D(j)} x_{ij} \leq c_j, \quad \forall j \in \{1, \ldots, n_2\}.$$

We again need to define constraints that enforce stability by ruling out the existence of any blocking pair. The following inequalities make sure that no blocking pair can be present in a solution:

$$c_j \Big(1 - \sum_{q \in H_j^{\leq}(i)} x_{iq} \Big) \leq \sum_{p \in D_i^{\leq}(j)} x_{pj}, \quad \forall i \in \{1, \ldots, n_1\}, \quad \forall j \in H(i).$$

The above constraints are the adaptation of the stability constraints in the previous problem when capacity is considered. More specifically, they ensure that if doctor i was not assigned to hospital j or any other hospital, they rank at the same level or higher than j (i.e., $\sum_{q \in H_j^{\leq}(i)} x_{iq} = 0$); then hospital j has filled its capacity with doctors it ranks at the same level or higher than i (i.e., $\sum_{p \in D_i^{\leq}(j)} x_{pj} = c_j$).

2.19 ▪ Maximum Number of Disjoint Paths in a Graph

(a) In the given network set the capacity of each arc to 1 unit and find the maximum flow from s to t. The maximal flow value gives you the number of arc-disjoint directed paths from s to t. If the maximum flow value is at least 2 units, then we can conclude that there exist at least two such paths from s to t.

(b) Since we have maximum flow = capacity of minimal capacity cut, we have that the maximum number of arc-disjoint paths in a graph is equal to the number of arcs in a minimal cardinality cut. This is known as Menger's theorem (1927).

2.20 ▪ Matrix Rounding and Maximum Flow

This problem is adapted from Ahuja et al. [1].[8]

---

[8]The authors of [1] adapted this problem from [3].

We form a directed graph as follows: we create a column of nodes for each row and a column of nodes for each column. Also we create two artificial nodes, the source and the sink. We draw an arc from the source to the "row" node i with flow lower bound equal to $\lfloor a_i \rfloor$ and upper bound equal to $\lceil a_i \rceil$. We draw an arc from "column" node j to the sink with flow lower bound equal to $\lfloor b_j \rfloor$ and upper bound equal to $\lceil b_j \rceil$. We draw an arc between row node i and column node j with lower bound equal to $\lfloor d_{ij} \rfloor$ and upper bound equal to $\lceil d_{ij} \rceil$. Finally, we draw an arc from the sink back to the source with infinite capacity (see Figure 15), and we look for the maximum integer flow from s to t. Since all lower and upper bounds are integers, and this is a network flow problem, an integer maximum flow always exists.

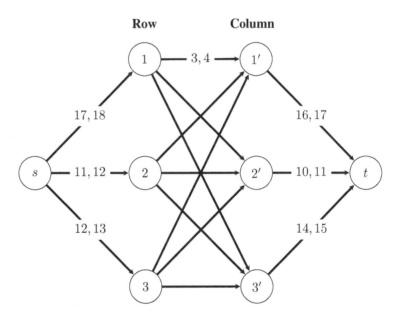

Figure 15. *Network for Matrix Rounding Example.*

Chapter 3

Studying Integer Optimization Problems

3.1 ▪ Integer Points in 2D

(a) The graph of $\mathcal{S}$ is shown in Figure 16.

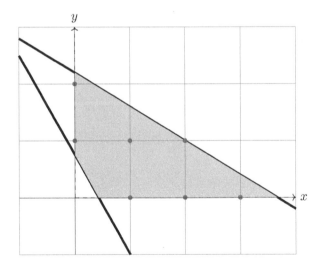

Figure 16. *Graph of the Region* $\mathcal{S}$.

(b) The points of $\mathcal{S} \cap \mathbb{Z}^2$ are $(0,1), (0,2), (1,0), (1,1), (2,0), (2,1), (3,0)$.

(c) The convex hull of $\mathcal{S} \cap \mathbb{Z}^2$ can be found by writing appropriate line equations and choosing the type of inequalities correctly.

$$\mathcal{P} = \left\{ (x,y) \in \mathbb{R}_+^2 : \begin{array}{c} x + y \geq 1, \\ x + 2y \leq 4, \\ x + y \leq 3. \end{array} \right\}$$

3.2 ▪ A Binary Knapsack

(a) The order of ratios is $4 \to 3 \to 2 \to 1$. LP relaxation gives $(0, 1, 1, 1)$, and since this is integer, we have found the solution.

(b) Let $I_1 = \{\mathbf{x} \in \mathbb{B}^4 : 2x_1 + x_2 + x_3 + x_4 \le 3\}$ and $I_2 = \{\mathbf{x} \in \mathbb{B}^4 : 97x_1 + 32x_2 + 25x_3 + 20x_4 \le 134\}$.

$$40x_1 + 20x_2 + 20x_3 + 20x_4 \le 60,$$
$$5x_3 \le 5,$$
$$12x_2 \le 12,$$
$$57x_1 \le 57.$$

Adding these together gives $97x_1 + 32x_2 + 25x_3 + 20x_4 \le 134$. Therefore if $\mathbf{x} \in I_1$, then $\mathbf{x} \in I_2$. Conversely, if $x_1 = 0$ for some $\mathbf{x} \in I_2$, we have $32x_2 + 25x_3 + 20x_4 \le 134$, which implies $x_2 + x_3 + x_4 \le 3$ and $2x_1 + x_2 + x_3 + x_4 \le 3$. Otherwise if $x_1 = 1$, then $32x_2 + 25x_3 + 20x_4 \le 37$. Hence at most one can be 1; therefore $2x_1 + x_2 + x_3 + x_4 \le 3$ holds and $I_2 \subseteq I_1$. Together we have $I_1 = I_2$, and $(0, 1, 1, 1)$ solves (P_2).

3.3 ▪ Hierarchy among LP Relaxations

This problem is adapted from a similar problem in [15].

(a) The three formulations are equivalent since all three sets consist of the points

$$(1, 1, 0, 0), (1, 0, 1, 0), (1, 0, 0, 1), (0, 1, 1, 1)$$

and all binary 4-element vectors that are dominated (componentwise smaller than) by at least one of these points.

(b) Let us explore the relationship between $LP(S_2)$ and $LP(S_1)$ first. If $x \in S_2$, then

$$32(2x_1 + x_2 + x_3 + x_4) + 33x_1 \le 32(3) + 33,$$

or equivalently

$$97x_1 + 32x_2 + 32x_3 + 32x_4 \le 129,$$

which dominates the inequality defining S_1, so $LP(S_2) \subseteq LP(S_1)$. The inclusion is strict. This can be seen by taking the point $(\frac{62}{97}, 1, 1, 1) \in LP(S_1)$, which is obviously not an element of $LP(S_2)$.

Similarly, if $x \in LP(S_3)$, then

$$\frac{37}{2}(x_1 + x_2 + x_3) + \frac{13}{2}(x_1 + x_3 + x_4) + \frac{27}{2}(x_1 + x_2 + x_4) + \frac{117}{2}x_1$$
$$\leq \frac{37}{2}(2) + \frac{13}{2}(2) + \frac{27}{2}(2) + \frac{117}{2},$$

or equivalently

$$97x_1 + 32x_2 + 25x_3 + 20x_4 \leq \frac{271}{2},$$

which dominates the inequality defining S_1, so $LP(S_3) \subseteq LP(S_1)$. Again, the inclusion is strict since $(\frac{62}{97}, 1, 1, 1) \in LP(S_1)$ is obviously not an element of $LP(S_3)$.

(c) We have

$$\mathbf{x} = \left(\frac{1}{2}, 1, 1, 0\right) \in LP(S_2) \setminus LP(S_3)$$

while

$$\mathbf{y} = \left(\frac{1}{2}, \frac{3}{4}, \frac{3}{4}, \frac{3}{4}\right) \in LP(S_3) \setminus LP(S_2),$$

so neither set contains the other. Therefore, a hierarchy cannot be established between the two.

3.4 ▪ LP Relaxations and Unboundedness

(a)

$$\begin{aligned} \text{maximize} \quad & 2x + 3y \\ \text{subject to} \quad & x + y \geq 3, \\ & x, y \in \mathbb{Z}. \end{aligned}$$

(b)

$$\begin{aligned} \text{maximize} \quad & x \\ \text{subject to} \quad & x = \sqrt{2}y, \\ & x, y \in \mathbb{Z}. \end{aligned}$$

3.5 ▪ Vertices of an Integral Polyhedron

We have to solve $x^2 + y^2 = 50$ for the points on the boundary, which are $(\pm 1, \pm 7)$, $(\pm 7, \pm 1)$, and $(\pm 5, \pm 5)$. An initial guess at the ideal formulation is that the polyhedron wraps these points. But one still has to check whether one has integer solutions between the polyhedron and the circle. By symmetry there are only two cases. Above the line between $(-1, 7), (1, 7)$ there may be integer solutions, but $(0, 7)$ is on the polyhedron and $(0, 8)$ is outside the circle, so we are safe. Above the line between $(1, 7), (5, 5)$ there may be integer solutions, but there are no integers between $(2, 6.5)$ and $(2, \sqrt{41})$. $(3, 6)$ is on the polyhedron, $(3, 7)$ is outside the circle, and the $(4, 5.5)$ and $(4, \sqrt{34})$ interval does not contain any integers. Therefore our initial guess provides an ideal formulation. Since the objective has all positive coefficients, the optimal solution must be one of $(1, 7), (7, 1), (5, 5)$, and it is $(5, 5)$ with value 25.

3.6 ▪ Branch-and-Bound Fails

(a) The objective function is lower bounded by 1, and the LP solution will always try to push x_3 to 1. But there are no integer pairs x_1, x_2 such that $12x_1 + 9x_2 = 1$ since $\gcd(12, 9) = 3 > 1$. Therefore, the branching process will take forever.

(b) If you add $x_3 \geq 3$ to the problem, even the LP returns a fractional solution, and Branch-and-Bound will eventually return an integral solution. That is, $(1, -1, 3)^T$.

3.7 ▪ Covers and Binary Knapsack

(a) The LP solution is $\left(1, 1, \frac{4}{7}, 0\right)^T$ with value $2 + \frac{4}{7}$.

(b) Set $C = \{1, 2, 3\}$ as a minimal cover with cover inequality $x_1 + x_2 + x_3 \leq 2$, which cuts the LP optimum.

(c) Include x_4 in the inequality, and obtain $x_1 + x_2 + x_3 + x_4 \leq 2$. This improvement cuts $\left(\frac{2}{3}, \frac{2}{3}, \frac{2}{3}, \frac{5}{27}\right)^T$.

3.8 ▪ Bin Packing Covers

Let us show the validity of the inequality using contradiction. Assume that the inequality is not valid. Then there exists a subset $C \subset \mathcal{N}$ such that

$$\sum_{i \in C} a_i > b,$$

and there exists a feasible solution (x, y) to the formulation of Section 1.2 satisfying

$$\sum_{i \in C} x_{ij} > (|C| - 1)y_j$$

for some $j \in \{1, \ldots, n\}$. We shall now consider two cases.

Case 1: If $y_j = 0$, i.e., the bin j is unused, for $\sum_{i \in C} x_{ij} > (|C| - 1)y_j$ to hold, we need x_{ij} to be equal to 1 for at least one $i \in C$, but such a solution does not exist due to the constraint

$$\sum_{i=1}^{n} a_i x_{ij} \leq by_j.$$

Case 2: If $y_j = 1$, i.e., the bin j is used, for $\sum\limits_{i \in C} x_{ij} > (|C| - 1)y_j$ to hold, we need x_{ij} to be equal to 1 for every $i \in C$, but such a solution is impossible due to the constraint

$$\sum_{i=1}^{n} a_i x_{ij} \le b y_j$$

and the nature of C, which satisfies $\sum\limits_{i \in C} a_i > b$.

Since we arrive at a contradiction in both cases, the inequality must be valid for all feasible solutions.

3.9 ▪ Integer Programming Solvers and Accuracy

The big-M constraint is $y \le 10^6 x$. Since $10^{-6} \sim 0$ according to the solver, the constraint will satisfy $(x, y) = (0, 1)$, which is a failure in computation.

3.10 ▪ A Quadratic Integer Program

There is no possibility for x_1 to be 0; hence we add the constraint $x_1 = 1$. Then, we conclude that $x_i = 1$ for all i. Therefore, we compute the optimal solution by plugging in the unique feasible solution, that is, $\sum\limits_{i=1}^{n} \sum\limits_{j=1}^{n} q_{ij} + \sum\limits_{i=1}^{n} p_i + R$.

3.11 ▪ Linearizing a Quadratic Integer Program

Define y_{ij} to be a binary variable for every pair i, j. Then we can write the following linear integer program, which is equivalent to the above quadratic 0-1 problem:

$$
\begin{aligned}
\text{minimize} \quad & \sum_{i=1}^{n} \sum_{j=1}^{n} q_{ij} y_{ij} + \sum_{i=1}^{n} c_i x_i \\
\text{subject to} \quad & y_{ij} \le x_i, && \forall i, j \in \{1, \ldots, n\}, \\
& y_{ij} \le x_j, && \forall i, j \in \{1, \ldots, n\}, \\
& y_{ij} \ge x_i + x_j - 1, && \forall i, j \in \{1, \ldots, n\}, \\
& \mathbf{x} \in \mathbb{B}^n, \\
& Y \in \mathbb{B}^{n \times n}.
\end{aligned}
$$

The above constraints make sure that $y_{ij} = x_i \cdot x_j$ for every i, j.

3.12 ▪ Set Covering

(a) Assume by contradiction that $\hat{\mathbf{x}}$ is optimal with $\hat{x}_j = 0$. Then we have $\mathbf{x} = \hat{\mathbf{x}} + \mathbf{e}_j$, which is feasible, and $\mathbf{c}^T \mathbf{x} = \mathbf{c}^T \hat{\mathbf{x}} + c_j < \mathbf{c}^T \hat{\mathbf{x}}$. Therefore $\hat{\mathbf{x}}$ cannot be optimal, and $x_j = 1$ for any optimal solution.

(b) If there is a row identical to $\mathbf{e}_j^T$, then we have $x_j \geq 1$ as a constraint, and binary restriction forces $x_j = 1$ for any feasible solution.

(c) Assume by contradiction that $\hat{\mathbf{x}}$ is optimal with $\hat{x}_j = 1$. Let $a_{ij} = 0$. Then for the ith constraint, $\mathbf{x} = \hat{\mathbf{x}} - \mathbf{e}_j + \mathbf{e}_k$ is feasible for such constraints. Let $a_{ij} = 1$. Then $a_{ik} = 1$; this makes $\mathbf{x} = \hat{\mathbf{x}} - \mathbf{e}_j + \mathbf{e}_k$ feasible for such constraints. Then

$$\mathbf{c}^T\mathbf{x} = \mathbf{c}^T\hat{\mathbf{x}} - c_j + c_k < \mathbf{c}^T\hat{\mathbf{x}},$$

and this shows that $\hat{\mathbf{x}}$ cannot be optimal. So $x_j = 0$ for any optimal solution.

3.13 ▪ Magic Squares Revisited

Suppose we have the linear relaxation (LP). Since M_n is an eigenvalue of S_n with the eigenvector $\mathbf{1}$, $\hat{\mathbf{x}} = \mathbf{1}$ is a feasible solution for (LP). Then we have the dual of the relaxation,

$$\begin{aligned} \text{maximize} \quad & M_n\mathbf{1}^T\mathbf{y} \\ \text{subject to} \quad & S_n^T\mathbf{y} \leq \mathbf{1}. \end{aligned} \tag{D}$$

Then, if we can find a feasible $\hat{\mathbf{y}}$ such that $n = M_n\mathbf{1}^T\hat{\mathbf{y}}$ and $S_n^T\hat{\mathbf{y}} \leq \mathbf{1}$, we say that $\hat{\mathbf{x}}$ is optimal in (LP). But taking $\hat{\mathbf{y}} = M_n^{-1}\mathbf{1}$ we satisfy both conditions, and we have $\hat{\mathbf{x}}$, an optimal solution for (LP). Since (LP) has a larger feasible region than (P) (in the sense of inclusion) and $\hat{\mathbf{x}}$ is an integral vector, we conclude that $\hat{\mathbf{x}}$ is an optimal solution of (P).

3.14 ▪ Independent or Stable Sets

(a) Assume by contradiction that $x_{i*} = 1$. Then $x_l = 0$ for any $l \in V \setminus \{i\}$ since there are $|V| - 1$ constraints containing x_{i*}. But if there are $j, k \in V$ such that $(j, k) \notin E$, then one can pick the feasible solution $\hat{\mathbf{x}}$ such that $\hat{x}_j = \hat{x}_k = 1$ with remaining entries zeros and $\sum_{i \in V} x_i < \sum_{i \in V} \hat{x}_i$. This contradicts the fact that $\mathbf{x}$ is optimal. Therefore $x_{i*} = 0$.

(b) If $|V|^2 - 2|E| \leq |V|$, we have $|E| \geq \frac{|V|(|V|-1)}{2}$, but the RHS is the maximum number of constraints possible in an (MISP), so we have $|E| = \frac{|V|(|V|-1)}{2}$. Then for any $i, j \in V$, we must have that $(i, j) \in E$. Hence for any optimal solution $\hat{\mathbf{x}}$, there is at most one k such that $\hat{x}_k = 1$ and rest are zeros. There are $|V|$ such optimal solutions, with the unique non-optimal feasible solution $\mathbf{x} = \mathbf{0}$.

(c) If $\mathbf{1}^T\mathbf{a}_i = 2$ for any $i \in V$, all variables appear in the constraints exactly twice. If we add all the constraints, we obtain

$$2\sum_{i \in V} x_i = |E| \Rightarrow \sum_{i \in V} x_i \leq \left\lfloor \frac{|E|}{2} \right\rfloor.$$

3.15 ▪ Chvátal–Gomory Cuts

(a) One can pick $u_1, u_2 = 0$ and $u_3 = \frac{7}{64}$; then one can see that

$$\left\lfloor \frac{-56}{64} \right\rfloor x_1 + \left\lfloor \frac{-63}{64} \right\rfloor x_2 \le \left\lfloor \frac{-7}{2} \right\rfloor \Rightarrow -x_1 - x_2 \le -4.$$

(b) For this part, we are looking for $-x_1 \le -2$, and we are assumed to find u_1, u_2, u_3 such that

$$0 > 7u_1 - u_2 - 8u_3 \ge -1,$$

$$1 > u_1 + 3u_2 - 9u_3 \ge 0,$$

$$-1 > 28u_1 + 7u_2 - 32u_3 \ge -2.$$

Let $A = \begin{bmatrix} 7 & -1 & -8 \\ 1 & 3 & -9 \\ 28 & 7 & -32 \end{bmatrix}$; then $A^{-1} = \begin{bmatrix} -\frac{3}{55} & -\frac{8}{55} & \frac{3}{55} \\ -\frac{4}{11} & 0 & \frac{1}{11} \\ -\frac{7}{55} & -\frac{7}{55} & \frac{2}{55} \end{bmatrix}$. Then we pick $\lambda_1 \in [-1, 0)$, $\lambda_2 \in [0, 1)$, and $\lambda_3 \in [-2, -1)$. Then $\hat{u}_1, \hat{u}_2, \hat{u}_3$ are uniquely determined by $A^{-1}\lambda$:

$$u_1 = \frac{1}{55}(-3\lambda_1 - 8\lambda_2 + 3\lambda_3) \in \left(\frac{-14}{55}, 0 \right),$$

$$u_2 = \frac{1}{11}(-4\lambda_1 + \lambda_3) \in \left(-\frac{2}{11}, \frac{3}{11} \right),$$

$$u_3 = \frac{1}{55}(-7\lambda_1 - 7\lambda_2 + 2\lambda_3) \in \left(-\frac{1}{5}, \frac{1}{11} \right).$$

The interval for u_1 contradicts the fact that $u_1 \ge 0$. Hence it is not possible to obtain $x_1 \ge 2$ using the given inequalities with the Chvátal–Gomory technique.

3.16 ▪ Extended Covers

(a) The extended cover inequality is $\sum_{i=1}^{m} x_i \le k - 1$ since $|C| = k$.

(b) Assume we multiply the first constraint with $\frac{1}{a_k}$, and for $x_i \le 1$'s we multiply them with $\frac{a_k - a_i}{a_k}$ and then add them up:

$$\sum_{i=1}^{k-1} x_i + \sum_{i=k}^{m} \frac{a_i}{a_k} x_i + \sum_{i=m+1}^{n} \frac{a_i}{a_k} \le \frac{b - \sum_{i=1}^{k-1} a_i}{a_k} + k - 1.$$

Here for $i \in \{k, \ldots, m\}$ we have $\frac{a_i}{a_k} \ge 1$, so $\left\lfloor \frac{a_i}{a_k} \right\rfloor \ge 1$ too. For $i \in \{m + 1, \ldots, n\}$ we have $a_i < a_k$; hence $\left\lfloor \frac{a_i}{a_k} \right\rfloor = 0$. For the RHS, $b - \sum_{i=1}^{k-1} a_i < a_k$;

hence $\lfloor \text{RHS} \rfloor = k - 1$. Therefore we obtain

$$\sum_{i=1}^{k} x_i + \sum_{i=k+1}^{m} \left\lfloor \frac{a_i}{a_k} \right\rfloor x_i \leq k - 1,$$

which may be a better inequality depending on the weights of items $\{k+1, \ldots, m\}$.

3.17 ▪ Diagonal Distances

Let $(x, y) \in S \cap \mathbb{Z}^2$. Then we search for a feasible pair $(u, v) \in \mathbb{R}_+^2$ such that

$$\left. \begin{array}{l} (x, y) \\ (x + u, y + v) \\ (x - v, y + u) \\ (x + u - v, y + u + v) \end{array} \right\} \in S \cap \mathbb{Z}^2$$

in order to have the square s defined by these corner points lay in S. An equivalent linear program is as follows:

$$
\begin{aligned}
\text{maximize} \quad & \xi_1 + \xi_2 \\
\text{subject to} \quad & a_i x + b_i y \leq c_i, & \forall i \in \{1, \ldots, m\}, \\
& a_i(x + u) + b_i(y + v) \leq c_i, & \forall i \in \{1, \ldots, m\}, \\
& a_i(x - v) + b_i(y + u) \leq c_i, & \forall i \in \{1, \ldots, m\}, \\
& a_i(x + u - v) + b_i(y + u + v) \leq c_i, & \forall i \in \{1, \ldots, m\}, \\
& -\xi_1 \leq u - v \leq \xi_1, \\
& -\xi_2 \leq u + v \leq \xi_2, \\
& x, y \in \mathbb{Z}, \\
& u, v, \xi_1, \xi_2 \in \mathbb{R}_+.
\end{aligned}
$$

3.18 ▪ AND Operator

(a) $r \geq \sum_{i=1}^{n} x_i - (n-1)$ (I) and $r \leq x_i$ for all $i \in \{1, \ldots, n\}$ (II). (i) is satisfied by (II). (ii) is satisfied by (I). (iii) is satisfied by (II). (iv) is satisfied by (I).

(b) $r \geq \sum_{i=1}^{n} x_i - (n-1)$ (I) and $nr \leq \sum_{i=1}^{n} x_i$ (II). (i) is satisfied by (II). (ii) is satisfied by (I). (iii) is satisfied by (II). (iv) is satisfied by (I).

(c) Let $x_1 = 0$, $x_i = 1$ for $i \geq 2$ and $r = \frac{n-1}{n}$. This point is feasible for the second formulation and infeasible for the first formulation.

3.19 ▪ Handling Bounds

This problem is an adaptation of results presented by Witzig.[9]

(a) Since the constraints for the problem are box constraints, in an optimal solution x_i should be equal to u_i or l_i for linear programs. Therefore if $a_i > 0$, $x_i = l_i$ to minimize the objective. Similarly if $a_i < 0$, $x_i = u_i$ to minimize the objective. Then we can divide $a_i = a_i^+ - a_i^-$ into positive and negative parts. Then rearranging the terms will lead us to the closed form solution.

(b) Similar to the first part.

(c) If $a_{\min} > b$, then for any $\mathbf{x}$ in the box provided by inequalities $\mathbf{l} \leq \mathbf{x} \leq \mathbf{u}$, we have $\mathbf{a}^T\mathbf{x} \geq a_{\min} > b$ and this causes infeasibility.

(d) If $a_{\max} \leq b$, then for any x in the box provided by inequalities $\mathbf{l} \leq \mathbf{x} \leq \mathbf{u}$, we have $\mathbf{a}^T\mathbf{x} \leq a_{\max} \leq b$ and this causes redundancy.

(e) $\mathbf{a}^T\mathbf{x} - a_i x_i + a_i x_i \leq b$ iff $x_i \leq \frac{b - \mathbf{a}^T\mathbf{x} + a_i x_i}{a_i}$. Then we have $-a_i x_i \geq -a_i u_i$ since a_i is positive and $\mathbf{a}^T\mathbf{x} - a_i x_i \geq a_{\min} - a_i l_i$ using previous parts. This implies $x_i \leq \frac{b - a_{\min} + a_i l_i}{a_i}$, and the floor comes from the fact that x_i is an integer.

(f) $\mathbf{a}^T\mathbf{x} - a_i x_i + a_i x_i \leq b$ iff $x_i \geq \frac{b - \mathbf{a}^T\mathbf{x} + a_i x_i}{a_i}$. Then we have $-a_i x_i \geq -a_i l_i$ since $-a_i$ is postive and $\mathbf{a}^T\mathbf{x} - a_i x_i \geq a_{\min} - a_i u_i$ using previous parts. This implies $x_i \geq \frac{b - a_{\min} + a_i u_i}{a_i}$, and the ceiling comes from the fact that x_i is an integer.

3.20 ▪ Deep Learning and Neural Networks

This problem is adapted from Anderson et al. [2].

A mixed integer representation for $\mathrm{gr}(\mathrm{ReLU}; [l, u])$ is readily obtained by inspection as

$$y \geq x,$$
$$y \leq x - l(1 - z),$$
$$y \leq uz,$$
$$y \geq 0,$$
$$(x, y, z) \in \mathbb{R} \times \mathbb{R} \times \mathbb{B}.$$

3.21 ▪ Deep Neural Networks II

This problem is a continuation of the previous one, and is again adapted from Anderson et al. [2].

---

[9] https://www.researchgate.net/publication/337783650_Computational_Aspects_of_Infeasibility_Analysis_in_Mixed_Integer_Programming

(a) We simply use the result of the previous problem to get

$$y \geq f(x),$$
$$y \leq f(x) - m^-(1 - z),$$
$$y \leq m^+z,$$
$$y \geq 0,$$
$$(x, y, z) \in [L, U] \times \mathbb{R} \times \mathbb{B}.$$

(b) The mixed integer set is defined by the inequalities from part (a):

$$y \geq x_1 + x_2 - \tfrac{3}{2},$$
$$y \leq x_1 + x_2 - \tfrac{3}{2} + \tfrac{3}{2}(1 - z),$$
$$y \leq \tfrac{1}{2}z,$$
$$y \geq 0,$$
$$(\mathbf{x}, y, z) \in [0, 1]^2 \times \mathbb{R} \times \mathbb{B}.$$

Its LP relaxation set (just relax the binary nature of z) is defined by

$$y \geq x_1 + x_2 - \tfrac{3}{2},$$
$$y \leq x_1 + x_2 - \tfrac{3}{2} + \tfrac{3}{2}(1 - z),$$
$$y \leq \tfrac{1}{2}z,$$
$$y \geq 0,$$
$$(\mathbf{x}, y, z) \in [0, 1]^2 \times \mathbb{R} \times [0, 1].$$

The LP relaxation contains points that are not in the convex hull of integer feasible solutions; e.g., the point $(x, y, z) = ((1, 0), 0.25, 0.5)$ is an element of the second set (the LP relaxation). However, one can observe that $y \leq 0.5x_2$ is valid for all feasible solutions of the first (mixed integer) set. To see this, set $z = 1$, in which case y can be equal to at most 0.5 when both $x_1 = x_2 = 1$, and when $z = 0$, we also have $y = 0$. So, at all points of the mixed integer set we have $y \leq 0.5x_2$. However, this inequality is violated by $(x, y, z) = ((1, 0), 0.25, 0.5)$.

3.22 ▪ Strenghtening Covers

(a) If $x_1 = 0$, then α_1 can be anything. So we look for the case when $x_1 = 1$. So we solve the following knapsack problem:

$$\begin{aligned}
\text{maximize} \quad & x_3 + x_4 + x_5 + x_6 \\
\text{subject to} \quad & 6x_2 + 6x_3 + 5x_4 + 5x_5 + 4x_6 + x_7 \leq 8, \\
& \mathbf{x} \in \mathbb{B}^7.
\end{aligned}$$

Then we see that $z^* = 1$; therefore the largest possible value for α_1 is $3 - z^* = 2$. Hence $c_1 : 2x_1 + x_3 + x_4 + x_5 + x_6 \leq 3$.

(b) We use a similar approach for α_2. Let $x_2 = 1$. So we solve the following knapsack problem:

$$\begin{aligned}
\text{maximize} \quad & 2x_1 + x_3 + x_4 + x_5 + x_6 \\
\text{subject to} \quad & 11x_1 + 6x_3 + 5x_4 + 5x_5 + 4x_6 + x_7 \leq 13, \\
& \mathbf{x} \in \mathbb{B}^7.
\end{aligned}$$

Then we see that $z^* = 2$; therefore the largest possible value for α_2 is $3 - z^* = 1$. Hence $c_1 : 2x_1 + x_2 + x_3 + x_4 + x_5 + x_6 \leq 3$.

(c) Simply, $E(C) = \{1, 2, 3, 4, 5, 6\}$ and $c_3 : x_1 + x_2 + x_3 + x_4 + x_5 + x_6 \leq 3$.

(d) It is easy to see that c_2 is stronger than c_3 since

$$2x_1 + x_2 + x_3 + x_4 + x_5 + x_6 \leq 3 \Rightarrow x_1 + x_2 + x_3 + x_4 + x_5 + x_6$$
$$\leq 3 - x_1 \leq 3$$

for $\mathbf{x} \in [0, 1]^7$.

3.23 ▪ Lifting

Let $\{z_1, \ldots, z_6\}$ be binary variables defined as

$$z_i := \begin{cases} 1 & \text{if the } i\text{th item is in the cover,} \\ 0 & \text{otherwise.} \end{cases}$$

For C to be a cover we should have $\sum_{j \in C} w_j > b$. Since we work with integral coefficients we can have $\sum_{j \in C} w_j \geq b + 1$, or equivalently $\sum_{j=1}^{6} z_j w_j \geq b + 1$. Also we can write the objective function as $\sum_{j=1}^{6} (z_j \hat{x}_j - z_j) + 1$. Hence we have the following MIP:

$$\begin{aligned}
\text{maximize} \quad & -z_1 - z_2 - \tfrac{1}{4}z_3 - \tfrac{1}{2}z_4 - 0z_5 - z_6 + 1 \\
\text{subject to} \quad & 45z_1 + 46z_2 + 79z_3 + 54z_4 + 53z_5 + 125z_6 \geq 179, \\
& \mathbf{z} \in \mathbb{B}^6.
\end{aligned}$$

The optimal solution is $\mathbf{z}^* = \begin{bmatrix} 0 & 0 & 1 & 1 & 1 & 0 \end{bmatrix}^T$. Then we see that $\hat{x}$ is cut by the inequality $x_3 + x_4 + x_5 \leq 2$.

3.24 ▪ Wheels and Tours

In order to have a Hamiltonian tour, we should have a solution satisfying the 2-matching constraint. Then node 0 can only use two edges. These should be connected to two consecutive nodes; otherwise it is impossible to find a tour covering all nodes. Therefore we have to make a decision on nodes $\{1, \ldots, n\}$. If a feasible solution has the connections $(0, i)$ and $(0, i+1)$, automatically it will avoid $(i, i+1)$; otherwise we will have a sub-tour. Therefore it suffices to add distances of the chosen connections to and subtract the avoided edge from our objective function because all edges of the type $(j, j+1)$ will be a part of a feasible tour for $i \neq j$.

3.25 ▪ On Shortest Paths

(a) For each arc in the graph we have a 1 and a -1 in the matrix A. This makes $2|E|$ non-zero elements. Therefore the number of zeros is $(|V| - 2)|E|$.

(b) Let $\mathbf{x}$ be a feasible solution. Then we have

$$\sum_{e \in E} A_{i^*,(e)} x_e = 0 \Rightarrow \sum_{e \in \mathcal{E}} x_e = 0 \Rightarrow x_e = 0, \quad \forall e \in \mathcal{E}.$$

(c) Assume there exists a matrix C such that $AC = I_{|V|}$. Then if we multiply both sides by a vector of ones,

$$\mathbf{0}^T = \mathbf{1}^T AC = \mathbf{1}^T I_{|V|} = \mathbf{1}^T,$$

which is a contradiction.

3.26 ▪ Deriving a Cutting Plane from Covering Constraints

This problem is taken from MATP6620/ISYE6760 Combinatorial Optimization & Integer Programming by J.W. Mitchell.[10]

Any binary vector that violates

$$\sum_{i=1}^{n} x_i \geq n - 1$$

must have at least two components with value 0, say $x_p = x_q = 0$, with $1 \leq p < q \leq n$. Then this point violates the corresponding constraint $x_p + x_q \geq 1$. This shows that the given inequality is satisfied by all solutions of the system

$$x_i + x_j \geq 1, \quad \text{for} \quad 1 \leq i < j \leq n.$$

Taking $x_i = 0.5$ satisfies the system but violates the given inequality.

---

[10]Used by kind permission of the author. https://homepages.rpi.edu/~mitchj/matp6620

3.27 ▪ Mixed Integer Rounding (MIR)

This and the next two problems are adapted from O. Günlük's lecture slides.[11]

Recall that for a general mixed integer set

$$Q = \{v \in \mathbb{R}_+, y \in \mathbb{Z} : v + y \geq b\}$$

the inequality

$$v \geq f(\lceil b \rceil - y),$$

where $f = b - \lfloor b \rfloor$, is valid and is known as the MIR inequality. To see why this is valid for Q, let us distinguish two cases: 1. If $y \geq \lceil b \rceil$, we have $v \geq 0 \geq f(y - \lceil b \rceil)$. 2. If $y \leq \lfloor b \rfloor$, then we have

$$v \geq b - y = f + (\lfloor b \rfloor - y) \geq f + f(\lfloor b \rfloor - y) = f(\lceil b \rceil - x)$$

(the second inequality is due to the facts that $0 < f < 1$ and $y \leq \lfloor b \rfloor$, and the final equality is due to $1 + \lfloor b \rfloor = \lceil b \rceil$).

Applying this to the set x, we obtain

$$v \geq \frac{3}{10}(8 - y),$$

which is valid for X.

3.28 ▪ Mixed Integer Rounding: Relax and MIR

Let $\hat{a}_j = a_j - \lfloor a_j \rfloor$ and $\hat{b} = b - \lfloor b \rfloor$ (note that these fractional parts $\hat{a}_j$ and $\hat{b}$ are always non-negative and strictly smaller than one). We shall rewrite the inequality defining the set P^1 first as follows:

$$\sum_{j:c_j<0} c_j v_j + \sum_{j:c_j\geq 0} c_j v_j + \sum_{j:\hat{a}_j<\hat{b}} \hat{a}_j y_j + \sum_{j:\hat{a}_j\geq\hat{b}} \hat{a}_j y_j + \sum_{j\in I} \lfloor a_j \rfloor y_j \geq b.$$

Since all v_j's are non-negative valued, the first term on the left is non-positive. Hence, if we neglect that term, we get an expression on the left which is at least as large. We can also replace the fourth summation by

$$\sum_{\hat{a}_j\geq\hat{b}} y_j$$

since

$$\sum_{\hat{a}_j\geq\hat{b}} y_j \geq \sum_{\hat{a}_j\geq\hat{b}} \hat{a}_j y_j.$$

---

[11]http://www.princeton.edu/~aaa/Public/Teaching/ORF523/ORF523_Lec17_guest.pdf

Therefore, we can write the relaxed inequality

$$\sum_{c_j \geq 0} c_j v_j + \sum_{\hat{a}_j < \hat{b}} \hat{a}_j y_j + \sum_{\hat{a}_j \geq \hat{b}} y_j + \sum_{j \in I} \lfloor a_j \rfloor y_j \geq b,$$

which is valid for P^1. Since the first two terms in the above inequality are non-negative, and the third and fourth terms should be integer valued, take as v the terms

$$\sum_{c_j \geq 0} c_j v_j + \sum_{\hat{a}_j < \hat{b}} \hat{a}_j y_j$$

and as y the terms

$$\sum_{\hat{a}_j \geq \hat{b}} y_j + \sum_{j \in I} \lfloor a_j \rfloor y_j$$

and apply the MIR inequality from the solution of the previous problem. We get the following MIR cut:

$$\sum_{c_j \geq 0} c_j v_j + \sum_{\hat{a}_j < \hat{b}} \hat{a}_j y_j + \hat{b} \left(\sum_{\hat{a}_j \geq \hat{b}} y_j + \sum_{j \in I} \lfloor a_j \rfloor y_j \right) \geq \hat{b} \lceil b \rceil,$$

which is valid for P^1.

3.29 ▪ Relax and MIR: Multiple Constraints

Combine the equations

$$-\frac{1}{8}(-v - 4x - s_1 = -4) + \frac{1}{8}(-v + 4x - s_2 = 0)$$

to get

$$x + \frac{1}{8}s_1 - \frac{1}{8}s_2 = \frac{1}{2}.$$

Rewrite it as an inequality:

$$x + \frac{1}{8}s_1 - \frac{1}{8}s_2 \geq \frac{1}{2}.$$

Apply Relax and MIR (recall $s_2 \geq 0$) to get

$$\frac{1}{2}x + \frac{1}{8}s_1 \geq \frac{1}{2},$$

and substitute for s_1 to get

$$-\frac{1}{8}v \geq 0 \implies v \leq 0.$$

Combined with the non-negativity restriction on v, we have the desired conclusion.

3.30 ▪ Flow Cover Inequalities

Define a new variable z_i for each facility i as the total shipment sent from an open facility, i.e.,

$$z_i = \sum_{j=1}^{n} y_{ij}.$$

Also define the total demand $b = \sum_{j=1}^{n} d_j$. So we have the system

$$\sum_{i=1}^{m} z_i \leq b,$$
$$z_i \leq u_i x_i, \quad \forall i \in \{1, \ldots, m\},$$
$$\mathbf{z} \in \mathbb{R}_+^m,$$
$$\mathbf{x} \in \mathbb{B}^m,$$

to which the flow cover inequality can be applied.

3.31 ▪ Split Cuts

The following background information and problem are from Wolsey [26].[12]

The two hyperplanes $\{\mathbf{x} \in \mathbb{R}^2 : x_1 + 4x_2 = \frac{11}{4}\}$ and $\{\mathbf{x} \in \mathbb{R}^2 : x_1 + x_2 = \frac{5}{4}\}$ intersect at the point $(\frac{3}{4}, \frac{1}{2})^T$. Consider the cuts $x_2 \leq 0$ and $x_2 \geq 1$. Joining the points where these two cuts intersect P (the first hyperplane intersects $x_2 = 0$ at the point $(\frac{11}{4}, 0)^T$; the second hyperplane intersects $x_2 = 1$ at $(\frac{1}{4}, 1)^T$), we obtain the split cut $4x_1 + 10x_2 \geq 11$.

3.32 ▪ Packings and Integer Knapsack Cuts

This problem and the next are from Hooker [11].[13]

(a) Note that the first two terms $7x_1$, $5x_2$ cannot by themselves satisfy the inequality, even if x_1 and x_2 are set to their largest possible value of 3. To satisfy the inequality, one must have

$$4x_3 + 3x_4 \geq 42 - 7 \cdot 3 - 5 \cdot 3,$$

i.e.,

$$4x_3 + 3x_4 \geq 6,$$

---

[12]Reproduced with permission from John Wiley & Sons, Inc. Copyright © 2021 by John Wiley & Sons, Inc. All rights reserved.

[13]Reproduced with permission from Springer Nature. John N. Hooker, *Integrated Methods for Optimization*, second edition, pg. 27, 2012, Springer Nature.

which implies

$$x_3 + x_4 \geq \left\lceil \frac{6}{4} \right\rceil = 2.$$

This kind of reasoning leads to integer knapsack cuts. Here, $\{1, 2\}$ or x_1, x_2 is called a *packing*. It is a *maximal packing* because no proper superset is a packing.

(b) Note that $\{1, 3\}$, $\{1, 4\}$, and $\{2, 3, 4\}$ are also maximal packings. Then the inequalities are obtained using the same reasoning as in part **(a)**.

An interesting point to note: Non-maximal packings also give rise to integer knapsack cuts, and they may be non-redundant, i.e., they are not necessarily dominated by other maximal packing cuts. For example, the packing $\{2, 3\}$ produces the cut $x_1 + x_4 \geq 3$, which is not dominated by the previous cuts we discussed.

3.33 ▪ Strengthening Integer Knapsack Cuts

For **(a)**, we just repeat the technique we learned in the previous problem using the packing $\{2\}$, i.e., we set $x_2 = 3$, and we get $7x_1 + 4x_3 + 3x_4 \geq 27$. Then dividing both sides of the inequality by 7, and taking the ceiling of all coefficients, we obtain the inequality

$$x_1 + x_3 + x_4 \geq 4.$$

For **(b)**, notice that when $x_1 + x_3 + x_4 = 4$, the maximum of the LHS $7x_1 + 5x_2 + 4x_3 + 3x_4$ is equal to 40, attained by choosing $x_1 = 3, x_2 = 3, x_3 = 1, x_4 = 0$. Hence, this is not enough to obtain 42 at the RHS. Therefore, we must have at least 5 by summing x_1, x_3, and x_4.

3.34 ▪ Interval Matrices and Integral Polytopes

Consider the constraint matrix for the given example:

	x_0	x_1	x_2	x_3	x_4	x_5	x_6	x_7	x_8	x_9	x_{10}	x_{11}	x_{12}
4	1	1	0	0	1	0	0	0	0	0	0	0	0
5	1	1	0	0	0	1	0	0	0	0	0	0	0
6	1	1	0	0	0	0	1	0	0	0	0	0	0
7	1	0	1	0	0	0	0	1	0	0	0	0	0
8	1	0	1	0	0	0	0	0	1	0	0	0	0
9	1	0	1	0	0	0	0	0	0	1	0	0	0
10	1	0	0	1	0	0	0	0	0	0	1	0	0
11	1	0	0	1	0	0	0	0	0	0	0	1	0
12	1	0	0	1	0	0	0	0	0	0	0	0	1

The matrix given in the problem is simply observed to be an interval matrix. Since all data are integers, the result follows. For the general case, the result may be proven using mathematical induction on T and N.

3.35 • Integer Programming Duality: Lagrangian Relaxation

The introduction and problem are adapted from Fisher [8].

We have Table 3.

Table 3. *Solving the Lagrangian Dual.*

u	x_1	x_2	x_3	x_4	$Z_D(u)$
0	1	0	0	1	20
6	0	0	0	0	60
3	0	1	0	0	34
2	0	1	0	0	26
1	1	0	0	0	18
$\frac{1}{2}$	1	0	0	1	19
$\frac{3}{4}$	1	0	0	1	18.5

Clearly, the function $Z_D(u)$ attains its minimum value equal to 18 for $u = 1$. Hence, we know that $Z \leq 18$. However, if we solve the LP relaxation of P1, we also get 18 as an upper bound on Z. Therefore, the Lagrangian dual did not achieve a better (smaller) upper bound compared to LP relaxation. Could we do better with the Lagrangian dual? The answer is in the next problem.

It can be shown in general (although we do not do this here) that Z_D is a piecewise-linear convex function of the type shown in Figure 17.

3.36 • An Improved Lagrangian Dual

We report the value of $Z_D(v_1, v_2)$ for several values of (v_1, v_2) in Table 4. Notice that in the last line of the table we obtain 16, which must be the optimal value of the dual problem since the point we obtain, $(1, 0, 0, 0)$, is feasible in the original integer program and gives the value 16, which is a lower bound on Z. Since the relaxation

Table 4. *Solving the Lagrangian Dual II.*

v_1	v_2	x_1	x_2	x_3	x_4	$Z_D(v_1, v_2)$
0	0	1	1	0	0	26
13	0	0	0	0	1	17
11	0	1	0	0	0	16

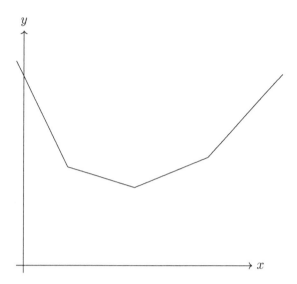

Figure 17. *The Lagrangian Dual Function.*

value cannot go lower than this value, we have found a provably optimal solution to the
integer problem. So, what happened? In the previous problem the upper bound from
the Lagrangian dual we found was equal to 18, which was the same as the LP relaxation
upper bound. The reason for this was that the Lagrangian subproblem

$$Z_D(u) = 10u + \text{maximize} \quad (16 - 8u)x_1 + (10 - 2u)x_2 - ux_3 + (4 - 4u)x_4$$
$$\text{subject to} \quad x_1 + x_2 \leq 1,$$
$$x_3 + x_4 \leq 1,$$
$$\mathbf{x} \in \mathbb{Z}_+^4$$

had the *integrality property*, which means that solving this problem as an LP always
yields an integer optimal solution. Whenever the Lagrangian subproblem has the integrality, we cannot do better than the LP relaxation. In contrast, the Lagrangian subproblem

$$Z_D(v_1, v_2) = v_1 + v_2$$
$$+ \text{maximize} \quad (16 - v_1)x_1 + (10 - v_1)x_2 - v_2x_3 + (4 - v_2)x_4$$
$$\text{subject to} \quad 8x_1 + 2x_2 + x_3 + 4x_4 \leq 10, \quad\quad (*)$$
$$\mathbf{x} \in \mathbb{Z}_+^4$$

does not have the integrality property (it is a knapsack problem!), and the bound im-
proved (in fact, in this example we found the optimal value Z).

Using subgradient optimization, for part **(c)** we obtain the values in Table 5 while
doing the iterations with a tolerance level $\epsilon = 10^{-4}$.

k	1	2	3	4	5	6	7	8	9	10	11	12	13	14	15	16	17
u_k	0	2	0	0.5	0.75	0.625	0.6875	0.7188	0.7344	0.7422	0.7461	0.748	0.749	0.7495	0.7498	0.7499	0.7499
t_l	1	0.5	0.25	0.125	0.0625	0.0312	0.0156	0.0078	0.0039	0.002	0.001	0.0005	0.0002	0.0001	0.0001	0	-
$(x_k)_1$	1	0	1	1	1	1	1	1	1	1	1	1	1	1	1	1	-
$(x_k)_2$	0	1	0	0	0	0	0	0	0	0	0	0	0	0	0	0	-
$(x_k)_3$	0	0	0	0	0	0	0	0	0	0	0	0	0	0	0	0	-
$(x_k)_4$	1	0	1	1	0	1	1	1	1	1	1	1	1	1	1	1	-

Table 5. *Subgradient Algortihm Iterations with $t_k = \frac{1}{2^k}$.*

k	1	2	3	4	5	6	7	8	9	10	11	12	13	14	15	16	17	18	19	20	21	22	23	24	25	26
u_k	0	2	0	1.5	0.75	0.375	0.9375	0.7031	0.8027	0.7024	0.7522	0.7287	0.7528	0.7411	0.7529	0.7471	0.75	0.7515	0.75	0.7507	0.75	0.7504	0.75	0.7502	0.75	0.7501
t_k	1	0.3125	0.75	0.0938	0.1875	0.2812	0.1172	0.0498	0.0502	0.0249	0.0118	0.0121	0.0059	0.0059	0.0029	0.0015	0.0007	0.0007	0.0004	0.0002	0.0002	0.0001	0.0001	0	-	-
$(x_k)_1$	1	0	1	0	1	1	1	1	1	1	1	1	1	1	1	1	1	1	1	1	1	1	1	1	1	-
$(x_k)_2$	0	1	0	1	0	0	0	0	0	0	0	0	0	0	0	0	0	0	0	0	0	0	0	0	0	-
$(x_k)_3$	0	0	0	0	0	0	0	0	0	0	0	0	0	0	0	0	0	0	0	0	0	0	0	0	0	-
$(x_k)_4$	1	0	1	0	1	0	1	0	1	0	1	0	1	0	1	0	1	0	1	0	1	0	1	0	1	-

Table 6. *Subgradient Algorithm Iterations with $t_k = \frac{\lambda_k(Z_D(u_k) - Z^*)}{\|b - Ax_k\|_2^2}$.*

This returns $Z_D(u_{17}) = 17.5$, which is slightly larger than what we obtained for $Z_D(v_1, v_2)$ but slightly smaller than the bound we obtained in Problem 3.35.

For part (**d**) we use the given update rule with the heuristic solution $\mathbf{x}_H = (1, 0, 0, 0)^T$ and $Z^* = 16$. Again tolerance ϵ is set to be 10^{-4}. See Table 6.

This returns $Z_D(u_{26}) = 17.5$, which is equal to what we obtained in part (**c**).

3.37 ▪ Simple Plant Location and Lagrangian Dual

Rewrite the Lagrangian subproblem

$$Z_D(\boldsymbol{\lambda}) = \sum_{i=1}^{n} \lambda_i + \quad \text{minimize} \quad \sum_{j=1}^{m} f_j y_j + \sum_{i=1}^{n} \sum_{j=1}^{m} (c_{ij} - \lambda_i) x_{ij}$$
$$\text{subject to} \quad x_{ij} \le y_j, \qquad \forall i \in \{1, \dots, n\}, \quad \forall j \in \{1, \dots, m\},$$
$$X \in \mathbb{B}^{n \times m},$$
$$\mathbf{y} \in \mathbb{B}^m,$$

which decomposes into independent subproblems for each facility j. We have thus $Z_D(\boldsymbol{\lambda}) = \sum_{j=1}^{m} Z_j(\boldsymbol{\lambda}) + \sum_{i=1}^{n} \lambda_i$, where for each facility site $j \in \{1, \dots, m\}$ we have

$$Z_j(\boldsymbol{\lambda}) = \quad \text{minimize} \quad f_j y_j + \sum_{i=1}^{n} (c_{ij} - \lambda_i) x_i^{(j)}$$
$$\text{subject to} \quad x_i^{(j)} \le y_j, \quad \forall i \in \{1, \dots, n\},$$
$$\mathbf{x}^{(j)} \in \mathbb{B}^n,$$
$$\mathbf{y} \in \mathbb{B}^m.$$

Each $Z_j(\boldsymbol{\lambda})$ can be computed easily. If $y_j = 0$, then $x_i^{(j)} = 0$ for all i, and the objective value is 0. If $y_j = 1$, all clients that are profitable are served, namely those with $c_{ij} - \lambda_i > 0$. Then the objective function value is $\sum_{i=1}^{n} \min[0, c_{ij} - \lambda_i] + f_j$. Therefore, we have $Z_j(\lambda) = \min\left\{0, \sum\limits_{i=1}^{n} \min[0, c_{ij} - \lambda_i] + f_j\right\}$. So we have $Z_D(\lambda) = \sum\limits_{i=1}^{n} \lambda_i + \sum\limits_{j=1}^{m} \min\left\{0, \sum\limits_{i=1}^{n} \min[c_{ij} - \lambda_i, 0] + f_j\right\}$.

3.38 ▪ A Dice Puzzle Solved by Dynamic Programming

Define the following function:

$$f_i(j) = \begin{cases} \text{number of ways to obtain } j \text{ by throwing } i \text{ dice,} & j \in \{1, n \times k\}, \\ 0 & \text{otherwise.} \end{cases}$$

So we set the boundary conditions by

$$f_1(j) = \begin{cases} 1 & \text{if } j \in \{1, \ldots, k\}, \\ 0 & \text{otherwise.} \end{cases}$$

We can define the update rule by

$$f_i(j) = \sum_{t=1}^{k} f_{i-1}(j - t).$$

Then we have to find $f_n(m)$ to compute the desired number. For $n = 3, k = 6$, and $m = 10$ we have $f_1(\{1, 2, 3, 4, 5, 6\}) = 1$ and 0 otherwise. To compute $f_3(10)$ we should know $f_2(\{9, 8, 7, 6, 5, 4\})$; hence

$$f_2(4) = f_1(-2) + f_1(-1) + f_1(0) + f_1(1) + f_1(2) + f_1(3) = 3,$$
$$f_2(5) = f_1(-1) + f_1(0) + f_1(1) + f_1(2) + f_1(3) + f_1(4) = 4,$$
$$f_2(6) = f_1(0) + f_1(1) + f_1(2) + f_1(3) + f_1(4) + f_1(5) = 5,$$
$$f_2(7) = f_1(1) + f_1(2) + f_1(3) + f_1(4) + f_1(5) + f_1(6) = 6,$$
$$f_2(8) = f_1(2) + f_1(3) + f_1(4) + f_1(5) + f_1(6) + f_1(7) = 5,$$
$$f_2(9) = f_1(3) + f_1(4) + f_1(5) + f_1(6) + f_1(7) + f_1(8) = 4.$$

Using these results, we obtain that

$$f_3(10) = f_2(4) + f_2(5) + f_2(6) + f_2(7) + f_2(8) + f_2(9) = 27.$$

3.39 ▪ Minimum Cut versus Maximum Cut

Unfortunately, passing from min to max in the formulation above does not work! Let us try to formulate Maximum Cut from scratch. Define $x_u = 1$ for $u \in V_1$ and $x_v = 0$ for $v \in V_2$. Since this takes care of a partition and all possible partitions are allowed, we do not have constraints. The problem is how to write the objective function. For every edge (u, v) if x_u and x_v have different values, then this is a contribution of 1 (or c_{uv} in the weighted case) to the objective function value. If we were allowed non-linear expressions, we could simply write $(x_u - x_v)^2$ and add this up for all edges in E. However, we cannot. Define a binary variable y_{uv} for each pair of vertices u, v.

Goal: Write a set of linear constraints that will force y_{uv} to take the value of the product $x_u \cdot x_v$ in all possible cases.

$$y_{uv} \leq x_u, \quad y_{uv} \leq x_v, \quad 1 - x_u - x_v + y_{uv} \geq 0.$$

Then we can write the objective function as $x_u + x_v - 2y_{uv}$. This is in fact exactly equal to $(x_u^2 + x_v^2 - 2x_u x_v)$ because for binary x_u (and x_v) we have $x_u^2 = x_u$. Hence, we arrive at the linear integer programming formulation

$$
\begin{aligned}
\text{maximize} \quad & \sum_{u,v \in V} x_u + x_v - 2y_{uv} \\
\text{subject to} \quad & y_{uv} \leq x_u, && \forall u, v \in V, \\
& y_{uv} \leq x_v, && \forall u, v \in V, \\
& 1 - x_u - x_v + y_{uv} \geq 0, && \forall u, v \in V, \\
& Y \in \mathbb{B}^{|V| \times |V|}, \\
& \mathbf{x} \in \mathbb{B}^{|V|}.
\end{aligned}
$$

3.40 ▪ The Cutting Stock Problem and Column Generation

This problem is adapted from [5].

(a) We have to find an upperbound on the objective initially and may choose $p = 610$ by inspection. In that case we have the following model:

$$
\begin{aligned}
\text{minimize} \quad & \sum_{j=1}^{610} y_j \\
\text{subject to} \quad & 135z_{1j} + 108z_{2j} + 93z_{3j} + 42z_{4j} \leq 300y_j, \quad \forall j \in \{1, \ldots, 610\}, \\
& \sum_{j=1}^{610} z_{1j} \geq 97, \\
& \sum_{j=1}^{610} z_{2j} \geq 610, \\
& \sum_{j=1}^{610} z_{3j} \geq 395, \\
& \sum_{j=1}^{610} z_{4j} \geq 211, \\
& Z \in \mathbb{Z}_+^{4 \times 610}, \\
& \mathbf{y} \in \mathbb{B}^{610}.
\end{aligned}
$$

After solving the problem above, we see that 453 rolls were needed to satisfy the given demand.

(b) Let us start with a set of solutions:

$$S' = \{(2,0,0,0), (1,1,0,1), (1,0,1,1)\}.$$

Iterations will be as follows:

Iteration No.	Dual optimal value	$\tilde{\mathbf{u}}$	Knapsack optimal value	$\mathbf{s}^*$
1	1005	$(0,1,1,0)$	3	$(0,0,3,0)$
2	741.6666667	$(0,1,\frac{1}{3},0)$	2	$(0,2,0,2)$
3	485.1666667	$(\frac{1}{2},\frac{1}{2},\frac{1}{3},0)$	1.1666	$(0,1,2,0)$
4	452.25	$(\frac{1}{2},\frac{1}{2},\frac{1}{4},0)$	1	$(0,1,2,0)$

The optimal solution for the relaxation consists of using 48.5 instances of pattern $(2,0,0,0)$, 206.25 instances of pattern $(0,2,0,2)$, and 197.5 instances of pattern $(0,1,2,0)$ with the objective 452.25. We can round this solution in order to obtain an integer solution. Demand constraints for the first and third items are active, so we cannot round down 48.5 and 206.25. If we have 49 instances of pattern $(2,0,0,0)$ and 207 instances of pattern $(0,2,0,2)$, then 197 instances of pattern $(0,1,2,0)$ would be sufficient to satisfy the demand. Hence we found 453 as the optimal value, as we did in part **(a)**.

Bibliography

[1] R. Ahuja, J. B. Orlin, and T. Magnanti, *Network Flows: Theory, Algorithms and Applications*, Prentice Hall, Englewood Cliffs, NJ, 1993.

[2] R. Anderson, J. Huchette, W. Ma, C. Tjandraatmadja, and J. P. Vielma, Strong mixed-integer programming formulations for trained neural networks, *Math. Program. Ser. B*, vol. 183, 3–39, 2020.

[3] M. Bacharach, Matrix rounding problems, *Management Science*, vol. 12(9), 627–744, 1966.

[4] P. S. Bradley and O. L. Mangasarian, Feature selection via concave minimization and support vector machines. ICML, 98:82–90, 1998.

[5] G. Gornuéjols, M. Conforti, and G. Zambelli, *Integer Programming*, Graduate Texts in Mathematics, Springer-Verlag, New York, 2014.

[6] M. Delorme, S. Garcia, J. Gondzio, J. Kalcsics, D. Manlove, and W. Pettersson, Mathematical models for stable matching problems with ties and incomplete lists, *European Journal of Operational Research*, vol. 277(2), 426–441, 2019.

[7] M. Fischetti, *Lezioni di Ricerca Operativa*, second edition, Edizioni Libreria Progetto Padova, Padova, Italy, 1999.

[8] M. Fisher, An applications oriented guide to Lagrangian relaxation, *Interfaces*, vol. 15(2): 10–21, 1985.

[9] C. Guéret, C. Prins, and M. Sevaux, *Optimization Applications with XPress-MP*, Editions Eyrolles, Paris, France, 2000.

[10] M. Helmling, S. Ruzika, and A. Tanatmış, Mathematical programming decoding of binary linear codes: Theory and algorithms, *IEEE Transactions on Information Theory*, vol. 58(7), 47–53, 2012.

[11] J. N. Hooker, *Integrated Methods for Optimization*, second edition, Springer, New York, 2012.

[12] M. Kocvara, Truss topology design with integer variables made easy, *Optimization Online*, 2010, https://optimization-online.org/?p=11135.

[13] G. Lancia, Integer programming models for computational biology problems, *J. Computer Sci. and Tech.*, vol. 19(1), 60–77, 2004.

[14] S. Maldonado, J. Pérez, R. Weber, and M. Labbé, Feature selection for support vector machines via mixed integer linear programming, *Information Sciences*, 279:163–175, 2014.

[15] G. Nemhauser and L. Wolsey, *Integer Programming and Combinatorial Optimization*, Wiley, New York, 1989.

[16] R. Phillips, *Pricing and Revenue Optimization*, Stanford Business Books, Stanford, CA, 2005.

[17] A. E. Roth, U. G. Rothblum, and J. H. Vande Vate, Stable matchings, optimal assignments, and linear programming, *Math. Oper. Res.*, vol. 18, 803–828, 1993.

[18] S. K. Sen, H. Agarwal, and S. Sen, Chemical equation balancing: An integer programming approach, *Mathematical and Computer Modelling*, vol. 44(7–8), 678–691, 2006.

[19] J. Shapiro, *Mathematical Programming*, Wiley, New York, 1977.

[20] G. Sierksma and Y. Zwols, *Linear and Integer Optimization: Theory and Practice*, third edition, CRC Press, Boca Raton, FL, 2015.

[21] G. Sierskma and D. Ghosh, *Networks in Action: Texts and Computer Problems in Network Optimization*, International Series in Operations Research and Management Science, Springer, New York, 2010.

[22] H. A. Taha, *Operations Research*, eighth edition, Pearson, Upper Saddle River, NJ, 2007.

[23] R. C. Walker, *Introduction to Mathematical Programming*, Prentice-Hall, Englewood Cliffs, NJ, 1999.

[24] H. P. Williams, *Model Building in Mathematical Programming*, fifth edition, Wiley, New York, 2013.

[25] W. L. Winston, *Operations Research: Applications and Algorithms*, fourth edition, Cengage Learning, Boston, MA, 2003.

[26] L. Wolsey, *Integer Programming*, second edition, Wiley, New York, 2021.

Index